Lecture Notes in Electrical Engineering

Volume 797

The book series *Lecture Notes in Electrical Engineering* (LNEE) publishes the latest developments in Electrical Engineering - quickly, informally and in high quality. While original research reported in proceedings and monographs has traditionally formed the core of LNEE, we also encourage authors to submit books devoted to supporting student education and professional training in the various fields and applications areas of electrical engineering. The series cover classical and emerging topics concerning:

- Communication Engineering, Information Theory and Networks
- Electronics Engineering and Microelectronics
- Signal, Image and Speech Processing
- Wireless and Mobile Communication
- Circuits and Systems
- Energy Systems, Power Electronics and Electrical Machines
- Electro-optical Engineering
- Instrumentation Engineering
- Avionics Engineering
- Control Systems
- Internet-of-Things and Cybersecurity
- Biomedical Devices, MEMS and NEMS

For general information about this book series, comments or suggestions, please contact leontina.dicecco@springer.com.

To submit a proposal or request further information, please contact the Publishing Editor in your country:

China

Jasmine Dou, Editor (jasmine.dou@springer.com)

India, Japan, Rest of Asia

Swati Meherishi, Editorial Director (Swati.Meherishi@springer.com)

Southeast Asia, Australia, New Zealand

Ramesh Nath Premnath, Editor (ramesh.premnath@springernature.com)

USA, Canada:

Michael Luby, Senior Editor (michael.luby@springer.com)

All other Countries:

Leontina Di Cecco, Senior Editor (leontina.dicecco@springer.com)

**** This series is indexed by EI Compendex and Scopus databases. ****

More information about this series at http://www.springer.com/series/7818

Maode Ma
Editor

Proceedings of the 4th International Conference on Telecommunications and Communication Engineering

ICTCE 2020, 4–6 December, Singapore

Springer

Editor
Maode Ma
School of Electrical and Electronic
Engineering
Nanyang Technological University
Singapore, Singapore

ISSN 1876-1100 ISSN 1876-1119 (electronic)
Lecture Notes in Electrical Engineering
ISBN 978-981-16-5691-0 ISBN 978-981-16-5692-7 (eBook)
https://doi.org/10.1007/978-981-16-5692-7

This Springer imprint is published by the registered company Springer Nature Singapore Pte Ltd.
The registered company address is: 152 Beach Road, #21-01/04 Gateway East, Singapore 189721, Singapore

Contents

Efficient Digital Beamforming for Bluetooth 5.1 Using Software Defined Radio

Fabricio Toasa[1], Luis Tello-Oquendo[1], Carlos Ramiro Penafiel-Ojeda[1,2], Anibal Llanga-Vargas[2], and Giovanny Cuzco[1](✉)

[1] Universidad Nacional de Chimborazo, 060108 Riobamba, Ecuador
{fatoasa.fie,luis.tello,carlospenafiel,gcuzco}@unach.edu.ec
[2] Instituto de Telecomunicaciones y Aplicaciones Multimedia (ITEAM), Universitat Politecnica de Valencia (UPV), Valencia, Spain

Abstract. In this paper, the angle of arrival (AoA) estimation and the implementation of beamforming for the new Bluetooth Low Energy communication standard version 5.1 has been evaluated through Software- defined Radio using a linear array antenna synthesis. The system consists of a 4 × 1 uniform linear array (ULA), a radio frequency (RF) transceiver and an FPGA digital signal processing unit. The proposed system reduces implementation and computational costs. It can be also adjusted for different antenna configurations with a higher number of elements in the array. The linear array antennas designed works at 2.4 GHz ISM operating band. The experiments were carried out in an indoor environment. The experimental results show a high level of precision for the positioning system, an excellent level between the main and side lobes in the radiation patterns, Taylor's synthesis has been applied to reject the side lobes as well.

Keywords: Angle of arrival · Beamforming · Bluetooth low energy 5.1 · Software-defined Radio · Uniform linear array

1 Introduction

Beamforming technology is becoming a crucial point in modern wireless communication systems such as fifth-generation (5G), Long-term Evolution Advanced (LTE-A), and Wireless Local Area Network (WLAN). Efficient exploitation of beamforming reduces the power consumption of base stations and user equipment [1, 2], eliminating unwanted sources of interference and improving the signal-to-noise ratio (SNR) of received signals. The phases and amplitudes of each element of the antenna array are extracted to estimate the Direction of Arrival (DoA) and to calculate the vector of weights to direct the radiation pattern toward the estimated DoA. Beamforming techniques require a large amount of processing; applications must be performed in real-time and at very high-speed [3], where the cost of transmitters and receivers is a significant drawback.

This work was supported by Universidad Nacional de Chimborazo under the research project "Caracterizacion de un sistema de telecomunicaciones basadas en radio definida por software aplicada en tecnologias inalambricas emergentes".

M. Ma (Ed.): ICTCE 2020, LNEE 797, pp. 1–10, 2022.
https://doi.org/10.1007/978-981-16-5692-7_1

The analog beamformer (ABF) has the advantage of reducing the implementation cost. The radiation pattern is obtained through the use of phase shifters and power dividers, after the stage of the low noise amplifier (LNA) [4]. However, it presents non-linearity in the radio frequency (RF) components and a limited bandwidth [5]. On the other hand, the digital beamformer (DBF) has greater flexibility, adaptability, and higher performance, with the disadvantage that each antenna requires a transceiver [6], RF signals are digitized in baseband flows, in phase (I) and quadrature (Q). These signals contain the amplitude and phase information of the received signals. The DBF evaluates and weights the phases and amplitudes in the digital domain to form the desired radiation pattern. Hybrid beamforming represents a solution in Multiple Input Multiple Output (MIMO) systems, where system cost and flexibility are compensated by combining DBF and ABF in the RF domain [7]; however, hybrid beamforming is of practical interest in massive MIMO systems due to the fact of cost and efficiency. In [8], a solution is proposed to reduce hardware costs for massive MIMO systems by replacing phase shifters with RF switches. Meanwhile in [1, 2], DBF is performed using RF switches and software-defined radio (SDR). Localization is an essential application in 5G mobile technology; however, the global positioning system (GPS) does not achieve high-precision positioning indoors due to obstruction from buildings [9]. This fact has generated interest in developers and researchers in recent years, where different technologies have been used to provide better performance in indoor location services. Bluetooth Low Energy (BLE) is the technology with the highest demand in location services thanks to its low energy consumption, high demand, low cost, and precision [10]. Although specific real-time applications require location accuracy down to the centimeter level, current Bluetooth devices estimate the position between two devices using the Received Signal Strength Indicator (RSSI) [11]. To overcome this problem, the Bluetooth Special Interest Group (SIG) added an optional direction-finding capability in the Bluetooth Core Specification v5.1 using two different techniques for signal processing [12], the Angle of Arrival (AoA) on the receiver and the Angle of Departure (AoD) on the transmitter. Both methods require an antenna array; the most used are linear, rectangular, or a circular array; the elements can be or not uniformly distributed [13]. These techniques, in combination with distance estimation, determine the precise location of a device. However, in this paper, we will only focus on the AoA proposed by Bluetooth SIG. To date, very few studies have been proposed for the indoor positioning system using the AoA method incorporated in BLE 5.1. In [14], the accuracy of the AoA is compared to the positioning of the Wi-Fi transmitter using the Channel Status Information (CSI) through the RSSI. Studies such as [15, 16] present an empirical and experimental method for positioning the BLE transmitter using USRP radios, with a configuration of 2 and 4 antennas in the receiver, respectively. Indoor localization features a limited range of precision for angular detection, where precision estimates an error of 1–1.5 m, where precision within a few centimeters is still a challenge. This paper presents a promising approach to reducing hardware costs, combining the analog pre/postprocessing of multiple antennas with digital processing of a single-channel receiver. By doing so, we improved the detection of objects, devices, and people in indoor environments using the AoA proposed by Bluetooth SIG. Besides, we implement beamforming to improve the spectral efficiency, increasing the capacity and speed of the system, providing a higher SNR while minimizing the interference levels for the Bluetooth Core Specification v5.1. The remainder of this paper is organized as follows. Section 2 describes the system model. The structure of the antenna array and its results are presented in Sect. 3. Section 4 describes signal acquisition and processing. Section 5 presents the results of the digital beamformer. Finally, Sect. 6 concludes the article.

Table 1. Bluetooth direction finding signal with CTE.

Preamble	Access address	PDU	CRC	CTE
8 bits	32 bits	16–2056 bits	24 bits	16–160 ps

2 System Model

2.1 BLE v5.1 System Characteristics

Bluetooth Low Energy (BLE) v5.1 is a Personal Area Network (PAN), operating in the ISM unlicensed band in the frequency range of {2.4—2.4835} GHz. BLE specifies 40 RF channels, where each channel has a bandwidth (BW) of 2 MHz. Channels are classified into advertising channels and data channels. The advertising channels are responsible for device discovery, and they use the labels 37, 38, and 39 at the frequencies of 2.402, 2.426, and 2.48 GHz, respectively. The remaining 37 channels are used for the exchange of information using the Frequency Hopping Spread Spectrum (FHSS) technique to mitigate interference with the WLAN operating in the same frequency band [12]. Four transmission modes are established for the link layer: LE1M, LE2M, LE125K, L500K with a transmission rate of 1Mbps, 2 Mbps, 125 Kbps, and 500 Kbps, respectively; the first two correspond to the encoded modes and the latter two following the uncoded modes. The BLE v5.1 modulation is Gaussian Frequency Shift Keying (GFSK). In the AoA method, the transmitter device uses a single antenna; the receiving device is considered a locator, which contains multiple antennas that scan the signal and estimate the phase difference of the signal due to the path difference in each of the antennas in the array [18]. The receiving device takes samples in phase and amplitude (I/Q) of the signal while switching between the antennas in the array; this allows us to estimate the relative distance between two devices as illustrated in Fig. 1. The package structure at the physical layer is modified to support AoA. The frame of the uncoded packets is modified, as well as the packet data unit (PDU). This new field called Constant Tone Extension (CTE), comprised of a series of digital '1s', guarantees that the receiver receives a constant frequency unbleached (uncoded) (see Table 1). The CTE consists of a protection period of 4 ps where no operation is performed, a reference period of 8 ps, and lastly, switching and sampling slots that vary between 1 and 2 ps depending on the developer's design.

2.2 Architecture for the Digital Beamformer Using a Single Receiver

An efficient, economical, and complete system for digital beamforming is proposed for the Bluetooth v5.1 communication standard. The SDR platform for this project consists of the AD-FMCOMMS4-EBZ card and the Zedboard Zynq 7000 SoC. The system in the receiver consists of a 4 × 1 linear rectangular microstrip array antennas, using to detect the information that is sent from the SDR transmitter; the four signals are multiplexed in time function utilizing an RF switch, the signals are converted into streams of digital IQ data by the receiver's direct conversion string. Digital data is transferred to the digital processing unit, signal control for the RF switch, and demultiplexing of the digital data is provided by the programmable logic (PL) of the SoC. This configuration obtains four digital signals in amplitude and phase (I/Q), one corresponding to each antenna. The AoA is estimated, and therefore the weights vector that is multiplied to the four received signals to direct the beam towards the direction of the source, as illustrated in Fig. 2. The AD-FMCOMS4-EBZ card contains the AD9364 chip, which is an integrated RF device that contains a transceiver with 12 bits in DAC and ADC and works in the 47 MHz to 6 GHz band in transmission mode and from the 70 MHz to 6 GHz for receive mode. The SoC Zedboard evaluation card provides the logical hardware and allows the communication and programmability of the AD-FMCOMMMS4-EBZ radio card through the Dual ARM Cortex-A9 processor, and the hardware programmability of the Artix-7 FPGA. For controlling, monitoring, and generating signals in the Bluetooth 5.1 standard, the Libiio library is used to establish communication between the host computer and the SDR through. The Libiio interface on the host is available through the libiio client while on the integrated platform through the libiio server. The libiio server and client use the Ethernet protocol to establish communication, libiio allows the generation of C code that runs on the Dual ARM Cortex-A9 processor and allows interaction with the radio's I/O using programming code.

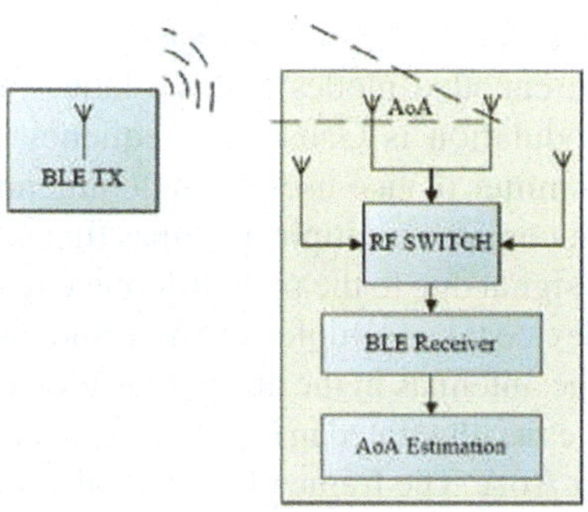

Fig. 1. Procedure to estimate the angle of arrival (AoA).

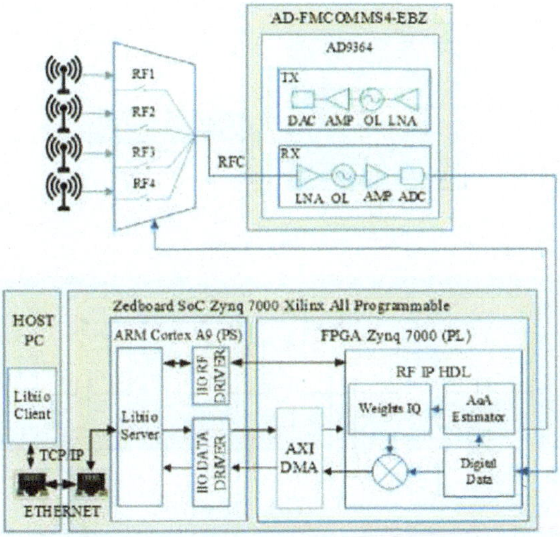

Fig. 2. Architecture for digital beamforming.

3 Uniform Linear Array Antenna Design

Due to its versatility and manufacturing cost, a rectangular patch microstrip array has been considered to demonstrate the system's operation. The antenna has been designed over a substrate FR-4, the parameters of the substrate are: Dielectric constant $e_r = 4.7$, thickness $h = 1.6$ mm, and loss tangent $S = 0.014$. The antenna has been designed and simulated in the CST Studio Suite electromagnetic software. The dimensions of a single element have been obtained according to the equations presented in [17]. The frontal and lateral view of the antenna are depicted in Fig. 3(a), and their optimized parameters to generate a proper matching at *50 Q* is shown in Table 2.

Table 2. Optimized parameters of single element microstrip antenna.

Parameters	Values [mm]	Parameters	Values [mm]
W and L	80	W_a	42.5
Lf	33.45	L_a	27.5
Wf	2.94	*Ix*	1.9

The array antenna has been designed based on the proposal presented in [19]. The design corresponds to a replica in 3 times along the horizontal axis, the distance that has been located between each isolated element is A/2 at the resonance frequency (2.4 GHz), i.e., a distance $d = 62.5$ mm. This separation, considered as a minimum value, will reduce the mutual coupling between the feeding ports. However, the side lobes' levels (SLL) are high, but they can be corrected through the synthesis of antennas. The prototype has been manufactured according to the specifications of the aforementioned FR-4 substrate.

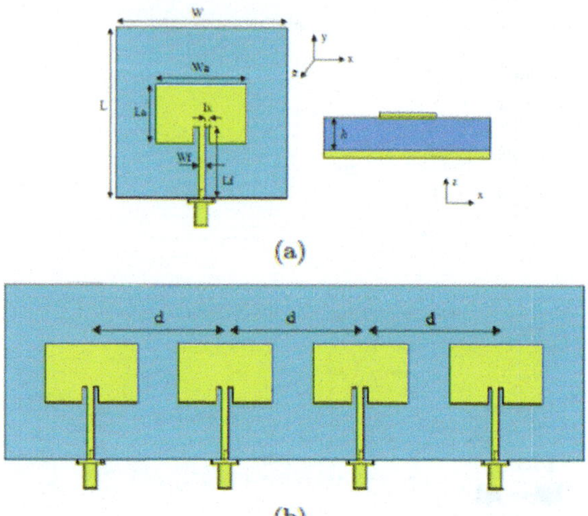

Fig. 3. Geometry of a rectangular microstrip antenna. (a) single element, and (b) uniform linear array of four elements at 2.4 GHz.

3.1 Array Antenna Results

A comparative plot of the simulated and measured reflection coefficient of the manufactured isolated antenna is depicted in Fig. 4(a). The matching band (Sn > −10 dB) extends from 2.402 GHz to 2.489 GHz. It is limited bandwidth; however, it is sufficient for the proposed application.

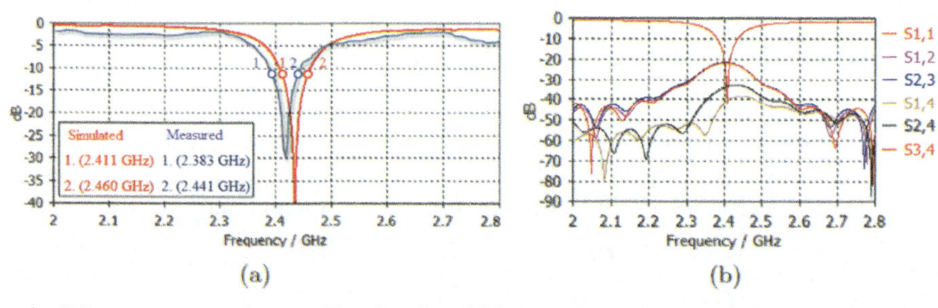

Fig. 4. S-Parameters results. (a) Simulated and Measured reaction coefficient, and (b) Mutual Coupling between the consecutive feeding ports.

The mutual coupling between the feeding ports has been depicted in Fig. 4(b); it is evident that the S parameters between the ports are very low, they are allless than −20 dB in the analysis frequency range. Lateral view of the simulated radiation pattern with a balanced power supply (the amplitude and phase values that enter each of the ports are the same) are illustrated in Fig. 5. Due to its configuration, it generates a directional broadside radiation pattern, the directivity generated by the structure is 12 dBi.

Fig. 5. Simulated radiation pattern at 2.4 GHz.

4 Acquisition and Processing of the Multiplexed Sign

The physical layer is configured for the Bluetooth version 5.1 standard by installing the Matlab support package for Radio based on Xilinx Zynq. The physical packets and CTE signals that are transmitted by the AD9364 chip transmitter are configured. The transmitter section is implemented in the FPGA programmable logic generated by Direct Digital Synthesizers (DDS). The Zed-board card is considered the baseband processor (BBP) for digital IQ signals; the signals are sent through the FPGA Mezzanine Connector (FMC) to the AD-FMCOMMS4-EBZ card. Advertising channel number 37 is chosen with a 2.402 GHz carrier for direction finding; the BLE waveform is transmitted through the air indefinitely.

4.1 Acquisition of Signals

The signals captured by the linear array of antennas are multiplexed by the RF switch HMC321LP4E, which has eight input signals and a standard RF port that connects to the receiver channel of the AD-FMCOMMS4 card. The programmable logic of the SoC defines the control of the RF switch. The FPGA programming file of the Matlab firmware is modified using Hardware Descriptive Language (HDL). Figure 6 indicates the RF switch used.

The receiver chain is a direct conversion system consisting of a Low Noise Amplifier (LNA), followed by down converter mixers that transform the RF signal into baseband, I/Q amplifiers, and Butterworth programmable 3 dB analog filters. Digitization is done using the 12-bit ADC converter. The digitized IQ signal is sent to the Base Band Processor of the SoC Zedboard through the 12-bit digital interface of the FMC connector for processing.

Fig. 6. Radiofrequency Switch HMC321LP4E.

4.2 Signal's Processing

The processing unit receives the digital I/Q signals. The receiver core in the FPGA is divided into two independent cores to support the sequence of I and Q samples. The samples validate the capture of the signal and are subjected to filtering in DC, performing a calibration for the I/Q imbalance in amplitude and phase mismatch. The corrected data is written to the external DDR memory through the Direct Memory Access (DMA) core through the AXI interface and is available in Matlab through libiio.

The data of the signal received in the radio hardware is presented as a complex matrix. Four columns of data are obtained where each column represents the complex data of each element of the linear array of antennas. The estimation of the AoA is performed using the Multiple Signal Classification (MUSIC) algorithm, considering the signal and noise subspaces to form the covariance matrix. Taylor's synthesis is established for beamforming. The algorithm reduces the responses of the elements of the antenna array and obtains a synthesized radiation diagram that satisfies the location of the main lobe, reduces the level of SLL, cancels projections to others directions, and improves the signal to interference ratio (SIR). The beamwidth is related to the SIR. The entire procedure is implemented in the signal processing unit, specifically in the programmable logic (PL). Figure 7 indicates the signal acquisition, switching, and processing stage implemented by the antenna array, RF switch, transceiver, and Zedboard SoC.

5 Experimentation and Results

The tests were conducted at the G-RESEARCH facilities of the National University of Chimborazo (UNACH). The transmitting antenna ('object') can be located at different unknown points in the range of [−90° to 90°] in the azimuth plane. In the tests, the object was located at 20°, the linear array of antennas with the SDR platform capture the transmitted signal, and employing the different algorithms implemented, the position of the object is detected, and the radiation pattern is aligned to the direction of interest.

A peak in the resulting pseudo spectrum represents the AoA; the MUSIC algorithm presents a high resolution. Figure 8 indicates the position of the transmitter when it is at 20°; the received power is normalized to its maximum level.

Fig. 7. Software-defined radio for digital beamforming.

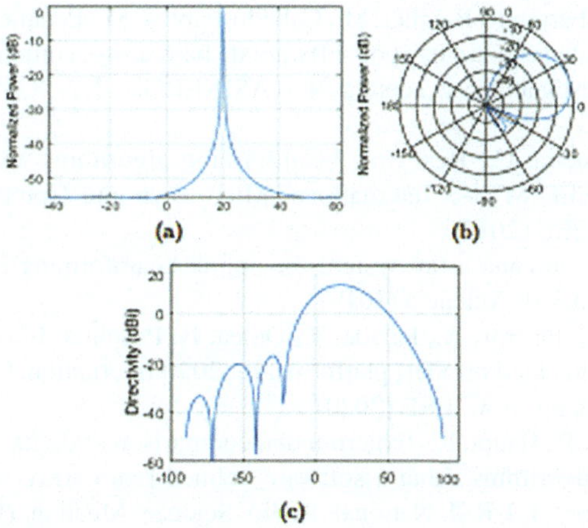

Fig. 8. Measured radiation pattern. (a) Angle of arrival at 20°, (c) Normalized power, and (d) Directivity at 2.4 GHz.

Taylor's synthesis adaptively weights phases and amplitudes for the antenna array signals. The calculated weights are multiplied to the digital data stored in the DDR memory; the following figures show the radiation pattern at 20° at the frequency of 2.4 GHz, with a ratio of main to secondary lobe at −28.65 dB, a beamwidth of 28°, with directivity of 11.75 dBi. By using a linear array of four antennas, the system supports a maximum inclination of the main beam in the range of [−45° to 45°] in the azimuth plane, high rejection of interferences, and placement of nulls in directions that do not correspond to the desired direction.

6 Conclusion

The implemented system provides an efficient and low-cost solution for digital beamforming for the Bluetooth Low Energy v5.1 communication standard using Software-defined Radio by the parallel processing of multiplexed signals. The positioning mechanism is validated by estimating the angle of arrival. It is observed that the resulting radiation pattern using a Taylor's synthesis supports a wide scanning angle, maintaining an excellent ratio between beamwidth and main-to-side lobe levels at −20 dB, high rejection of interference in the 2.4 GHz operating band by canceling side lobes. The implemented system could be used with higher antenna capacities for the new era of mobile 5G and wireless digital communication, applying the appropriate protocols.

References

1. Haroun, M.H., et al.: Sampled antenna array digital beamforming for LTE-advanced. In: International Conference on High Performance Computing & Simulation 2017 (HCPS), pp. 282–287. IEEE (2017)

2. Llanga-Vargas, A., Ferrando-Bataller, M., Cabedo-Fabres, M., Penafiel-Ojeda, C.R.: Sistema de agrupacion de antenas definidas por software de bajo costo, como instrumento de medida de MIMO, para investigacion y academia. NOVASINERGIA **1**(2), 83–89 (2018). ISSN 2631–2654
3. Chinatto, A., Junqueira, C.: Real time beamforming algorithms: experimental validation. In: 2017 SBMO/IEEE MTT-S International Microwave and Optoelectronics Conference (IMOC), pp. 1–5. IEEE (2017)
4. Haroun, M.H.: 8x1 antenna array system for uplink beamforming in LTE-A and 5G NR. Universitat Politecnica de Valencia (2019)
5. Codau, C., Buta, R., Pastrav, A., Palade, T., Dolea, P., Puschita, E.: An overview of digital beamforming implemented on SDR platforms. In: 2020 International Workshop on Antenna Technology (iWAT), pp. 1–4. IEEE (2020)
6. Gaydos, D., Nayeri, P., Haupt, R.: Experimental comparison of digital beamforming interference cancellation algorithms using a software defined radio array. In: 2019 United States National Committee of URSI National Radio Science Meeting (USNC-URSI NRSM), pp. 1–2. IEEE (2019)
7. Molisch, A.F., et al.: Hybrid beamforming for massive MIMO: a survey. IEEE Commun. Mag. **55**(9), 134–141 (2017)
8. Zhang, S., Guo, C., Wang, T., Zhang, W.: On-off analog beamforming for massive MIMO. IEEE Trans. Veh. Technol. **67**(5), 4113–4123 (2018)
9. Fan, M., Yang, F., Deng, Z., Jia, B., Leng, B.: A stochastic geometry approach to modeling time hopping based TDOA in 3D indoor localization. In: Proceedings of the 3rd International Conference on Telecommunications and Communication Engineering, pp. 130–135 (2019)
10. Marce, J.: Enhancing bluetooth location services with direction finding. In: Bluetooth Special Interest Group, Technical report, vol. 1 (2019)
11. Suryavanshi, N.B., Reddy, K.V., Chandrik, V.R.: Direction finding capability in bluetooth 51 standard. In: Kumar, N., Venkatesha Prasad, R. (eds.) International Conference on Ubiquitous Communications and Network Computing, vol. 276, pp. 53–65. Springer, Cham (2019). https://doi.org/10.1007/978-3-030-20615-4_4
12. Wooley, M.: Bluetooth core specification v5.1, January 2019
13. Manikas, A.: Beamforming: Sensor Signal Processing for Defence Applications. Imperial College Press (2015)
14. Tian, H., Zhu, L.: MIMO CSI-based super-resolution AoA estimation for Wi-Fi indoor localization. In: Proceedings of the 2020 12th International Conference on Machine Learning and Computing, pp. 457–461 (2020)
15. Cominelli, M., Patras, P., Gringoli, F.: Dead on arrival: an empirical study of the bluetooth 5.1 positioning system. In: Proceedings of the 13th International Workshop on Wireless Network Testbeds, Experimental Evaluation & Characterization, pp. 13–20 (2020)
16. Monfared, S., Nguyen, T.-H., Petrillo, L., De Doncker, P., Horlin, F.: Experimental demonstration of BLE transmitter positioning based on AOA estimation. In: 2018 IEEE 29th Annual International Symposium on Personal, Indoor and Mobile Radio Communications (PIMRC), pp. 856–859. IEEE (2018)
17. Hollander, D.: How AoA and AoD changed the direction of bluetooth location services. In: Bluetooth Blog, pp. 1–12 (2019)
18. Mailloux, R.J.: Phased Array Antenna Handbook. Artech House (2017)
19. Nasir, S.A., Mustaqim, M., Khawaja, B.A.: Antenna array for 5th generation 802.11 ac Wi-Fi applications. In: 2014 11th Annual High Capacity Optical Networks and Emerging/Enabling Technologies (Photonics for Energy), pp. 20–24. IEEE (2014)

Discussion on Suppression Coefficient of Projectile-Carried Communication Jamming

Jie Zhang[1(✉)], Haixia Shao[2], and Baohong Ma[3]

[1] Army Academy of Artillery and Air Defense, Hefei 230031, China
[2] 31441 Troops, Shenyang 110000, China
[3] Communication NCO Academy Army Engineering University, Chongqing 400035, China

Abstract. Focusing on the problem of jamming suppression coefficient in the design and development of projectile-carried communication jammer, this paper combed the existing references from three aspects: the concept of jamming suppression coefficient, the common values of jamming suppression coefficient, and the definition of jamming level. Finally, the value and definition of jamming suppression coefficient of projectile-carried communication were discussed through cases. The conclusion shows that in the actual design and development process of projectile-carried communication jammer, the jamming suppression coefficient does not have the same value range as the existing references, so the value of jamming suppression coefficient should be analyzed concretely according to the specific jamming scenarios.

Keywords: Projectile-carried communication jamming · Jamming suppression coefficient · Jamming level · Receiver modulation mode

1 Introduction

Projectile-carried communication jammer is an electronic device that uses artillery, rocket launcher and other launch equipment to carry it to the designated target area to conduct electronic jamming on enemy communication equipment [1, 2].

Compared with the traditional communication jammer, the projectile-carried communication jammer is obviously different from the traditional communication jammer in terms of equipment size space, launching power or transient flight time in the air. For example, the projectile-carried communication jammer is small, the launching power should not be too large, and the flying time in the air is short. Therefore, these limiting factors must be considered when designing and developing the projectile-carried communication jammer. According to the theory of communication jamming, we all hope that the smaller the jamming suppression coefficient is, the better it is for any kind of jamming, especially for projectile-carried communication jamming. For the same jamming object, relatively small jamming suppression coefficient will consume smaller jamming power, which is more favorable for the design and development of projectile-carried communication jamming prototype.

M. Ma (Ed.): ICTCE 2020, LNEE 797, pp. 11–17, 2022.
https://doi.org/10.1007/978-981-16-5692-7_2

However, the existing research and discussion on jamming suppression coefficient have different opinions in different references. Therefore, it is necessary to analyze and sort out the research contents of existing references on jamming suppression coefficient, so as to lay an early technical foundation for the design and development of projectile-carried communication jamming prototype.

2 The Concept of Jamming Suppression Coefficient

The definition of reference [3] is: when the effective jamming is reached, the ratio of the jamming power inside (or outside) the input end of the communication receiver to the signal power. The so-called effective jamming refers to the bit error rate p_e of digital communication or the error rate p_e of analog communication system reaching a certain level ε. When $p_e \geq \varepsilon$, it is said that the communication is effectively suppressed, otherwise it is not effectively suppressed.

The definition of reference [4] is the minimum jamming to signal ratio at the receiver input when the signal is effectively suppressed, that is, the minimum value of the ratio of jamming signal power to signal power.

3 Common Values of Jamming Suppression Coefficient

Reference [4]: in the case of jamming digital communication, if the bit error rate of digital signal reaches 20%, then for general digital communication system, the signal quality can not meet the requirements of performance index, and at the same time, the jamming effectively suppresses the signal.

Reference [4]: for analog voice communication, when the noise jamming power of the receiver output is 5–25 times of the voice signal power, the receiver can not correctly perceive the voice signal, and the jamming effectively suppresses the signal. That is, the suppression coefficient should be 5–25.

Reference [4]: in general, if the same frequency jamming is greater than or equal to the signal intensity, the jamming can work. The other kinds of jamming require that the jamming is much stronger than the signal. From the jamming power utilization rate, the same frequency jamming is undoubtedly the best. In other words, for the same frequency jamming, as long as the suppression coefficient is not less than 1.

References [5] and [6]: when the bit error rate is 2%, the communication information is usually unreliable. In order to ensure the reliability of information transmission, the information should be sent repeatedly. When the bit error rate reaches 50%, the digital information is completely disturbed, and the information can not be transmitted. The jamming completely suppresses the communication.

Reference [7]: the main communication mode of frequency hopping communication is frequency modulation voice communication. Generally, jamming power is required to be 10–20 dB higher than signal power to achieve better jamming effect.

Reference [8]: if the bit error rate can be increased to 0.1, then the jammer can jam the jamming target and achieve effective jamming.

According to reference [9], the jamming effect of frequency hopping communication is mainly determined by four parameters: frequency accuracy, jamming to signal ratio, channel jamming rate and jamming suppression time ratio, and the jamming to signal ratio usually needs 10–20 dB.

Reference values of suppression coefficient of common communication systems under normal white noise jamming are given in reference [3], as shown in Table 1:

Table 1. Reference values of suppression coefficient of common communication systems under normal white noise jamming

Communication system		Suppression coefficient
Voice communication ($\varepsilon = 0.5$)	DSB	3.2
	SSB	1.6
	VSB	1.6
	AM (Sine wave 100% modulation)	1.1
	Wideband FM (FM index is m_f)	4.8 $m_f^2(m_f + 1)$
Digital communication ($\varepsilon = 0.2$)	2ASK coherent detection	0.56
	2ASK incoherent detection	0.2
	2FSK coherent detection	0.7
	2FSK incoherent detection	0.55
	2PSK coherent detection	1.4
	2DPSK differential detection	1.1

When the jamming is effective, the bit error rate is not less than 0.2 for digital communication. For voice communication, the error rate of monosyllabic words is not less than 0.5.

4 Definition of Jamming Level

Reference [3]:
for digital communication, the jamming level can be divided into:

Strong jamming, $p_e = 0.2$;
Moderate jamming, $p_e = 0.12$;
Weak jamming, $p_e = 0.05$;

for voice communication,

Strong jamming, $p_e = 0.5$;
Moderate jamming, $p_e = 0.3$;
Weak jamming, $p_e = 0.1$.
References [3, 10, 11]:

In military affairs, the jamming level of digital communication system is divided as follows:

when $P_e \geq 0.2$, strong jamming, and the jamming level is three;
when $0.12 \leq P_e < 0.2$, moderate jamming, and the jamming level is two;
when $0.05 \leq P_e < 0.12$, weak jamming, and the jamming level is one;
when $P_e < 0.05$, it is unjammed.

According to the threat level and degree of jamming environment to communication system in reference [12], electromagnetic environment can be divided into 4 levels: 1, 2, 3, and 4. In addition, level 0 is defined as the electromagnetic environment without hostile jamming, as shown in Table 2:

5 Value and Definition of Suppression Coefficient of Projectile-Carried Communication Jamming

From the above analysis, it could be seen that the definition and value of communication jamming suppression coefficient (including jamming level) in different references were not the same, the difference was also large, and some of them were contradictory, but it was not to say who was right and who was wrong. How to define it depends on the specific jamming situation.

Here is an example [3, 13, 14]:

In analog voice communication, the lexical clarity and sentence intelligibility of voice are related to the signal-to-noise ratio $\rho_0 = \frac{S_0}{N_0}$ at the output of communication receiver when the pass-band is wide enough. The error rate P_e is related to ρ_0.

Suppose D1 is sentence intelligibility, D2 is the articulation of monosyllabic compound words, D3 is the articulation of monosyllabic words, and D4 is the articulation of meaningless syllables, their expressions are as follows:

$$
\begin{aligned}
D_1 &= 1 - \exp\left[-0.06128(\rho_0+12)^{1.6951}\right] \\
D_2 &= 1 - \exp\left[-0.03285(\rho_0+12)^{1.8319}\right] \\
D_3 &= 1 - \exp\left[-0.01741(\rho_0+12)^{1.6025}\right] \\
D_4 &= 1 - \exp\left[-0.01222(\rho_0+12)^{1.6708}\right]
\end{aligned}
$$

Where ρ_0 is expressed in decibels. Their relationship curves are shown in the Fig. 1:

In military affairs, voice communication quality can be classified as follows. Take D3 as an example,

when D3 ≤ 0.5, voice communication is seriously distorted;
when $0.5 <$ D3 ≤ 0.7, voice communication is moderately distorted;
when $0.7 <$ D3 ≤ 0.9, voice communication is weakly distorted;
when $0.9 <$ D3 ≤ 1, voice communication is not distorted.

Table 2. General classification standard of electromagnetic jamming environment

Electromagnetic environment level	0 level	1 level	2 level	3 level	4 level
Complexity	None	Simple	Weak	Moderate	Strong
Jamming mode	Unmanned jamming	Single frequency jamming	Single frequency jamming/tracking jamming	Single frequency jamming/tracking jamming/partial band jamming	Single frequency jamming/tracking jamming/partial band jamming/other complex jamming
Standard $\frac{S_0}{J_0}$ value	10 dB	5 dB	0 dB	−10 dB	−20 dB
Minimum $\frac{S_0}{J_0}$ value	7 dB	5 dB	−10 dB	−20 dB	−30 dB
$(ERP)_j/(ERP)_s$	0–3 dB	5–10 dB	10–20 dB	20–30 dB	30–40 dB
When the minimum $\frac{S_0}{J_0}$ of receiver detector is 10 dB, the required value of Gp	0–3 dB	5–10 dB	10–20 dB	20–30 dB	30–40 dB
N	0	1–3	4–8	9–18	19–40 or >40
W_j/W_c	0	No more than 5%	5%–10%	10%–20%	20%–50%

Among them:
$\frac{S_0}{J_0}$ _The power ratio of communication signal to jamming signal arriving at receiver antenna when meeting normal communication requirements;
$(ERP)_j$, $(ERP)_s$ _Average effective radiation power of jammer and receiver;
G_P _Processing gain of frequency hopping radio;
N _Number of jammed channels;
W_j/W_c _Percentage of jammed frequency band in communication frequency band.

Therefore, it can be seen from Fig. 1 that according to the requirements of different situations, the specific definition of the jamming level or jamming suppression coefficient for the military communication quality does not have a fixed value range, and the actual value should be analyzed in detail.

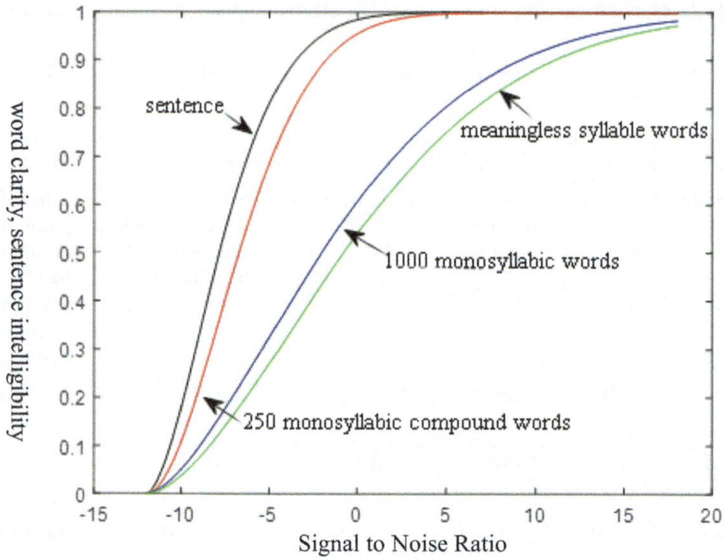

Fig. 1. The relationship between word clarity, sentence intelligibility and signal to noise ratio

6 Conclusions

By combing and analyzing the existing references, we can see that different scenarios need different considerations about the value of suppression coefficient of projectile-carried communication jamming. It is not like the references that there is a fixed value range, but the specific value should be analyzed according to the specific situation.

References

1. Ma, J., Hu, Z., Zhao, J.: Communication jamming projectile: a blinding weapon in the information battlefield. China's Adv. Weapons Rep.: Attack Defense (5), 65–67 (2005)
2. Yang, Y., Wang, D.: Bulgarian communication jamming projectile. Foreign Electron. Warfare (2), 38–42 (1998)
3. G\Shao, G., Cao, Z., He, J., et al.: EW Operational Effectiveness Analysis. PLA Press, Beijing (1998)
4. Wang, M., et al.: Communication Countermeasure Principle. PLA Press, Beijing (1999)
5. Hao, D.: Research on satellite communication jamming technology, p. 12. Xi'an University of Electronic Science and Technology (2012)
6. Zhao, H., Yang, Y., Wu, J., Feng, J.: Evaluation of jamming effect of LEO Satellite Communication. J. Sci. Instr. (S2), 1402–1405 (2006)
7. Kou, Z.: Research and simulation of jamming technology for frequency hopping communication system. Xi'an University of Electronic Science and Technology, pp. 2–3 (2010)
8. Na, D., Zhao, W.: Frequency hopping communication jamming and anti-jamming technology, p. 6. National Defense Industry Press, Beijing (2013)

9. Kou, Z.: Research and simulation of jamming technology for frequency hopping communication system, pp. 33–38. Xi'an University of Electronic Science and Technology (2010)
10. Li, B., Cai, X., Chen, H.: An evaluation model of digital communication jamming effect. Commun. Countermeas. (3), 14–15 (2005)
11. Zhang, R., He, M., Wang, Z.: Analysis and simulation of communication jamming bit error rate. Ship Electron. Countermeas. **31**(4), 73–76 (2008)
12. Yang, L.; Simulation and communication efficiency of frequency hopping radio in complex electromagnetic environment, p. 22. Central South University (2012)
13. Xu, Y.: Research on objective evaluation method of speech intelligibility, pp. 1–8. Taiyuan University of Technology (2015)
14. Xu, Y.: Research on objective evaluation method of speech intelligibility, pp. 39–45. Taiyuan University of Technology (2015)

Research of Key Technique of Simulation Test Grid for C4ISR System

Jing Jiang and Weibin Zhao(✉)

Tongda College of Nanjing University of Posts and Telecommunications, Yangzhou, China

Abstract. Based on the analysis of simulation test bed, the simulation test grid concepts for C^4ISR system is firstly proposed with the net-centric and service-oriented characteristics of new generation C^4ISR system into consideration. Then, the component, architecture, application model and implement principle of the simulation test grid are presented. Finally, with its application in a case, the simulation test grid is confirmed that it's an important method for the SoS (system of system) simulation.

Keywords: Test grid · Community of interest · Simulation test resource

1 Introduction

1.1 A Subsection Sample

The concept and technology of simulation-test-bed for distributed command and control system in the 90s of last century still have strong vitality [1]. It mainly meets the requirements of system design and development test by building system (components) in the loop simulation integration and test environment by using DIS or HLA framework [2].

The simulation test bed is a simulation test environment developed for specific tasks and system functions. It uses the test mode of the system-in-the-loop, and the function and structure of the simulation environment is relatively fixed. With the development of cloud computing technology, Li Bohu and so on proposed the cloud simulation concept [3–5], and built the cloud simulation platform, which is applied to the whole life cycle of the research and development, integration and operation of various distributed simulation systems [6]. Lu Shuang applies cloud simulation mode to the construction of air traffic management system, and solves the problem of multi-user/multi granularity resource sharing and collaboration [7].

With the development of a new generation of C4ISR systems in the direction of networking, service and integration, it has become a consensus [8, 9] to construct, run and organize the application of C4ISR based on the common military information infrastructure (military information grid). The new generation of C4ISR system can dynamically aggregate and use the resources on the network to adapt to the changes in the battlefield

M. Ma (Ed.): ICTCE 2020, LNEE 797, pp. 18–25, 2022.
https://doi.org/10.1007/978-981-16-5692-7_3

environment and tasks. The system boundary is fuzzy. It has the capability characteristics of the system function node with access, structure dynamic reconfiguration, function cooperative operation, information service-on-demand and so on.

Obviously, the network expands the boundary of the new generation system, and gives the system more resources, robust ability and flexibility [10]. It has put forward new application requirements for simulation application mode, simulation architecture, modeling and simulation operation, test evaluation and so on. The model of simulation test bed with relatively fixed function and structure is no longer available. It could not meet the application requirements of the new generation C4ISR system simulation test.

In order to support the design and development test of the new generation C4ISR system, this paper puts forward the concept of the simulation test grid on the basis of the simulation test bed technology research and application practice, and further describes the structure of the grid structure, the basic principle and the application prospect of the simulation test grid, and verifies the typical application cases.

2 Basic Concept and Connotation

The C4ISR simulation test grid is a test network composed of communication network, computing facility, C4ISR system simulation function unit, simulation test resource, simulation service and so on. It supports all kinds of simulation test functional unit access, simulation test resource sharing, joint cooperative operation, and provides simulation application design, development, operation and evaluation environment and tools for simulation personnel in the way of simulation portal, and organizes various test tasks by simulation task community [10].

The simulation test function unit includes the simulation entity model, the operable simulator, the simulation system, the real system (prototype)/equipment, etc. The simulation test resources include the simulation scenario library, the simulation model library, the software library, the basic database, the equipment performance library and the other data and model resources.

3 Simulation Grid Architecture

The C4ISR system simulation test grid is in essence the unified management and maintenance of all kinds of simulation facilities and test resources distributed on the network according to the unified specification and requirements. Under the unified scheduling and deployment of the simulation test management center, the simulation test resources are integrated dynamically according to the requirements of different simulation tests. Together, it forms one or more simulation task community to complete various levels and types of simulation test tasks. It is flexible, open, configurable and extensible.

The simulation grid architecture is shown in Fig. 1 which includes the infrastructure layer of simulation test, the core service layer of simulation test, the common service layer of simulation test and the application layer of simulation test.

The infrastructure layer of the simulation test provides the basic software and hardware operating environment and the simulation test data management for the upper level system simulation test, including communication equipment, network equipment, computing equipment, storage equipment, security protection equipment, database management system and so on.The core service layer is mainly to provide the core service support ability in the simulation grid environment, which has the similar technical features as the grid.The common service layer is based on the core service layer. It includes three parts. One is the simulation test management center, which is mainly responsible for the community management of the test task.The second part is all kinds of C4ISR analog resources, including real resources, virtual resources and simulation models. The third part is the common service of all kinds of simulation tests, such as the community configuration tool, the simulation test operation management tool, the data acquisition tool, the comprehensive situation display tool and so on. On the basis of the following three layers, the simulation test application layer forms a variety of simulation test task community through the dynamic integration of the simulation test resources, such as the architecture verification task community, the key technology verification task community, the system integration test task community and so on. Several test tasks could be completed at the same time in the test grid.

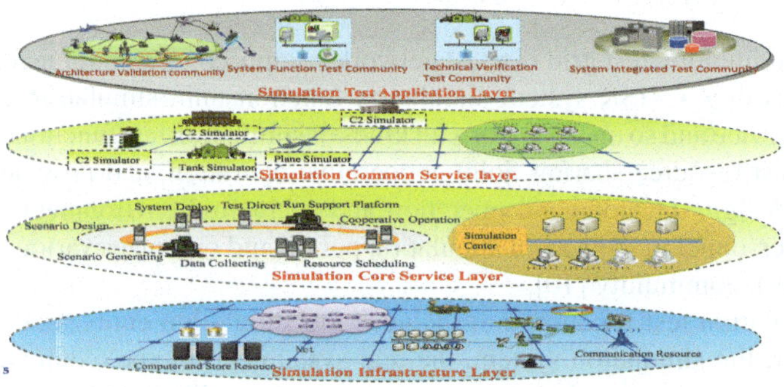

Fig. 1. Schematic diagram of the simulation grid architecture

4 Key Technology in Simulation Grid

4.1 Simulation Task Community Constructing Technology On-Demand

The construction of simulation task community is to integrate the existing simulation resources and construct service execution process according to certain business logic, so as to better meet the network simulation requirements.

According to the configuration mode of test system, it can be divided into static construction and dynamic construction. In the static configuration process, the test integration personnel bind the specific resources according to the test requirements, and set the property parameters of the experimental resources, including the entity number, the host number, the IP address, the system, the seat and so on. In the process of generating task community, the initialization information of each simulation resource will be generated according to configuration parameters and will not change.

In the process of dynamically configuring simulation task community, the test system integration personnel can choose simulation resources to participate in the simulation task community according to the requirement. For each simulation resource, in the dynamic configuration process, we only need to describe the attribute requirements and information interaction requirements of the simulation resources involved in the simulation task community, and do not specifically bind the attribute parameters to the instance objects. When the simulation task community is created, the simulation operating management center matches the test resource service according to the configuration attribute parameters, binding the specific test resources and dynamically generates instance objects. In the process of selecting test resources, multiple simulation resource instances often appear to meet the requirements, so it is necessary to optimize the selection of multiple resource services. At present, the optimization algorithm can be based on QoS optimization algorithm, such as particle swarm optimization or genetic algorithm.

4.2 Task Community Concurrent Operation Technology

In the process of constructing and running the test system, the simulation grid uses the integrated method based on the test service bus to build the supporting environment, and provides the discovery and integration function of the test resources, simulation models and tool services of various participating test task communities. It can integrate multiple simulation resources to support concurrent execution of multi-task communities.

When the test integrator completes the process of the community configuration of the test task, the test personnel can initiate the test request through the simulation grid portal to the test operation management center. After receiving the task community creation request, the operation management center will traverse the list of the configured test task community to determine whether the community configuration information exists. If existing, it will start to create the test task community dynamically. This process includes the selection and deployment of test resources and the core service. When the test system is deployed, the system will complete the configuration of the information interaction relation between different community members in the system based on the test bus, as shown in Fig. 2. The running instance of each task community is defined as the virtual executor. In different simulation virtual executor, it is necessary to create instances of simulation resources. For the same simulation resources, it may be subordinate to multiple simulation task communities, so the same simulation members also have multiple instances for different virtual executor services. In order to maintain the operation of the same task community and the correctness of information interaction, the management center will manage a multidimensional linked list. The management center reads all the registered test task community configuration information from the metadata directory and forms the first chain list: community 1 to N community, each task community has the unique identity μ. For each task community i, because there may be multiple instances, all the related virtual executor of the community i is associated with it, and the second level chain list is established. Each virtual executor j also has unique identities ν, and any virtual executor k can be identified by marking (μ, ν) in the running process of the management center. Each instance has a unique identity, which can receive and process request messages sent by the client members through a unique identity.

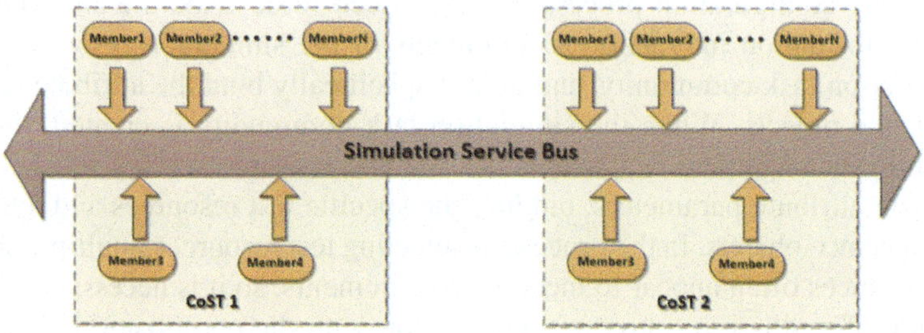

Fig. 2. Simulation service bus logic conceptual schematic

4.3 Information On-Demand Dispatch Technology

There are a large number of heterogeneous resources in the simulation grid, and the interaction information structure between any resources is different. According to the grammatical structure and concept connotation of the data items, the correlation and transformation rules between different fields in the heterogeneous interactive information are established to realize the conversion of the heterogeneous and interactive information in accordance with the demand. There are two ways of conversion: automatic conversion based meta-model and preset algorithm library. These two ways need to be described in advance in the process of defining the interactive metadata dictionary. As for the heterogeneity of some data items, such as units, precision, type and length, the preset transformation rules can be implemented, and the complex transformation of some data items can be realized by presetting conversion function in the configuration of interactive metadata dictionary. In order to ensure that the interactive information of different test resources can be understood and converted by the machine during the operation of the test community, the bus implements a transformation framework based on the information description meta-model, as shown in Fig. 2.

For any member of the test community, the metadata of the interactive information in the task community is provided in the construction phase of the task community, including the information structure description and the physical meaning of the data items. At the same time, when the interaction between members is established, the transformation rules of interactive information between two members are required to be configured. The configuration is based on the data item description information and the corresponding transformation rules based on the data items defined in the interactive metadata dictionary. After the establishment of a heterogeneous information conversion relationship between the two test members, the bus determines whether the conversion is needed in the course of the simulation task community according to the type of information sent/received (Fig. 3).

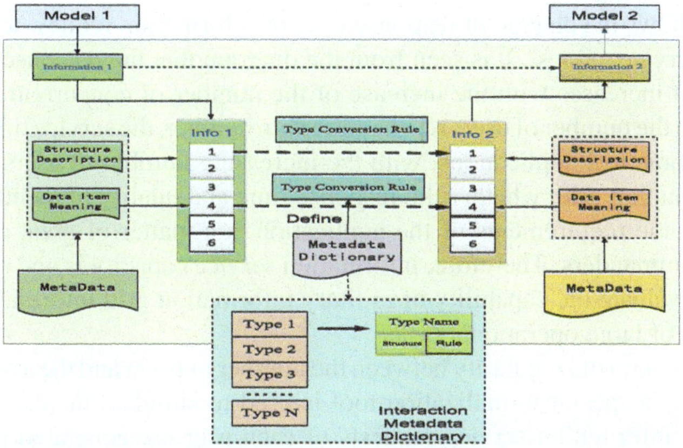

Fig. 3. Schematic diagram of dynamic transformation of interactive information

5 Typical Application Case

Taking the network system information service test and evaluation as an example, this paper introduces the system information on demand service capability test and evaluation method based on simulation test grid. As the simulation test technology is in the research stage, the described case is only a preliminary reflection of the research results. The information on demand service is mainly composed of information subscription service and information distribution service. Its ability involves the time, capacity and quality of information service. Table 1 is the main content of simulation test of information on demand service capability.

Table 1. Main content of simulation test of information on demand service capability.

Experiment subjects	Experiment content	Experiment content description
Service time	Information subscription average time	Relation of test subscription content, conditions and subscription service time
	Information dispatch average time	Relation of the count of users and the cost time of dispatch service
	Concurrent information subscription service time	Relation of the count of concurrent subscription services and the service time
Service capacity	Max of accessible users	The maximum number of users
Quality of service	Quality of subscriptions and dispatch services	The quality of information service

Figure 4 shows the change of response time of information request with concurrent information service request. It is seen from the diagram that the response time of information request increases with the increase of the number of concurrent users' service requests; when the number of users reaches a certain number, the service time is increased obviously. In practical applications, with the increasing number of users accessing the military information grid, whether the response time and quality of its information service can meet the requirements of the application is a matter of great concern to the operational commanders. Therefore, information service capacity is one of the effective indicators to evaluate the capability of military information grid information service to meet the scale of large operational tasks.

In order to obtain the regularity between the number of users and the average response time of service, a special virtualization tool is used to simulate the concurrent access users, and the information service requests of each user are generated randomly, and the experimental results as shown in Fig. 5. The graph shows that the average response time of information service increases with the number of users, showing a trend of curve growth. As the number of users increased from 10 to 100, the average time of information service increased from 1.541 s to 4.394 s. When the number of users is less than 60, the average response time of service increases slowly; when the number of users is more than 60, the average time of information service increases with the number of users.

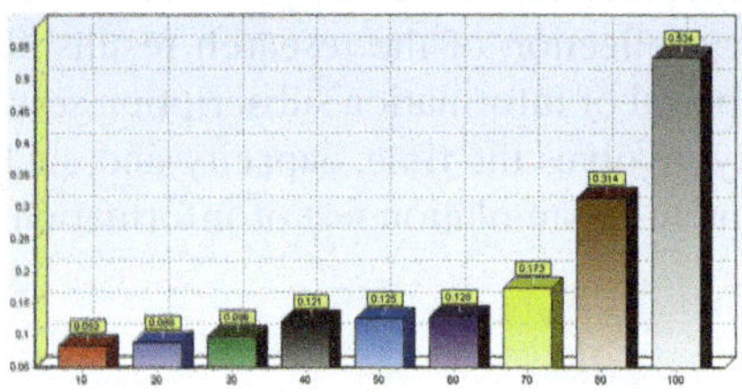

Fig. 4. The relationship between the number of concurrent information service requests and information service response time

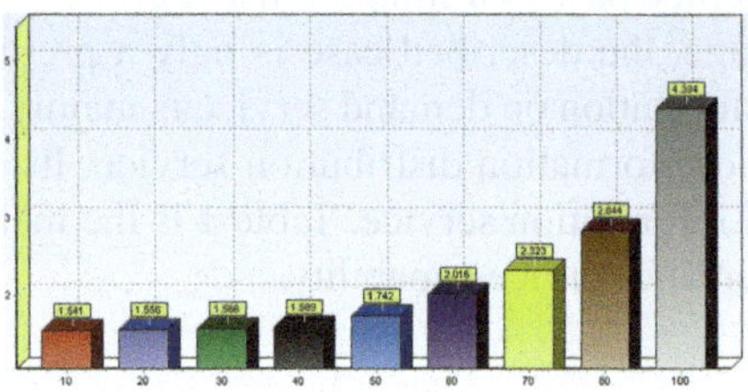

Fig. 5. The relationship between the count of users and the average respond time of service

In this experiment, the quality of information service is measured by the ratio of the number of information services satisfying to the user's requirement and the number of total information service. The results of the information service quality test are as shown in Fig. 4. It can be seen from the graph that the quality of information service decreases as the number of users increases. When the number of users is not more than 70, the results of the information service quality test fluctuate in 0.99. When the number of users is more than 70, the quality of information service is obviously reduced. When the number of users is equal to 100, the results of the information service quality test are reduced to about 0.90.

6 Conclusion

The traditional system-in-the-loop simulation test bed model has not adapted to the characteristics of the network and service of the new generation C4ISR system. The development and innovation of the simulation test grid technology is the supporting platform for the new generation of C4ISR system confrontation simulation test. The simulation portal, the simulation test capability is generated on demand, the simulation test object is connected with the access, the distributed heterogeneous joint simulation test and the cooperative operation are similar (Fig. 6).

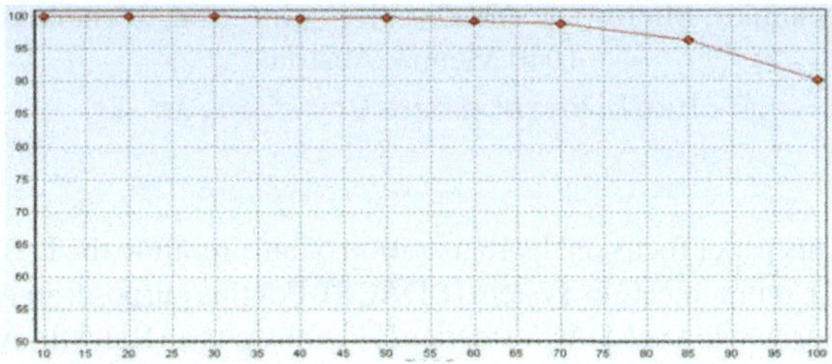

Fig. 6. The relationship between the quality of information service and the number of users

Acknowledgement. This work was Supported by Foundation of Tongda college of Nanjing University of Posts & Telecommunications (XK203XZ18013).

References

1. Zhang, C., Mao, S.: Using semi-practicality simulation method to construct simulation test bed of C3I information system. Comput. Simul. **19**(6), 100–102 (2002)
2. Li, B.: Networked modeling & simulation platform based on concept of cloud computing cloud simulation platform. J. Syst. Simul. **21**(17), 5292–5294 (2010)
3. He, H., Yuan, P., Cao, W.: Research on core technologies of grid-based distributed simulation platform. J. Syst. Simul. **2**(17), 437–442 (2014)
4. Li, B., Cai, X., Hou, B.: New Distributed Collaborative Simulation System-Simulation Grid. Jouranl of System Simulation **20**(20), 5423–5430 (2013)
5. Zhang, Y., Li, B., Chai, X., Yang, C.: Research on virtualization-based cloud simulation running environment dynamic building technology. Syst. Eng. Electron. **3**, 619–624 (2012)
6. Lu, S., Li, C.: Application conception of cloud simulation technology for ATM system. Command Inf. Syst. Technol. **8**(2), 71–76 (2017)
7. Xing, L., Chen, J.: A new method of simulation system construction. Comput. Mod. **12**, 72–76 (2013)
8. Mao, S., Li, Y., Lin, J.: Research on concepts and technique of network-centric simulation. J. Syst. Simul. **7**(22), 1660–1663 (2012)
9. Pullen, J.M., et al.: Using web services to integrate heterogeneous simulations in a grid environment. In: Bubak, M., van Albada, G.D., Sloot, P.M.A., Dongarra, J. (eds.) ICCS 2004. LNCS, vol. 3038, pp. 835–847. Springer, Heidelberg (2004). https://doi.org/10.1007/978-3-540-24688-6_108
10. Zhu, X.: Construction of data resource catalog system for equipment test. China CIO News **5**, 111–113 (2016)

Random Access Performance Evaluation and Improvements of the LDACS

Michael Zaisberger(✉) and Holger Arthaber

Institute of Electrodynamics, Microwave and Circuit Engineering, Technische Universitat Wien, 1040 Vienna, Austria
michael.zaisberger@tuwien.ac.at

Abstract. This paper focus on the registration of an aircraft to the L-Band Digital Aeronautical Communication System (LDACS). For this purpose, an Aircraft Station (AS) according to the LDACS standard [1] transmits an Random Access (RA) frame via one of two RA slots to an LDACS Ground Station (GS). Registration of an AS is typically required for two scenarios. One case is that an AS turns on its radio at the airport near to a GS, which corresponds to a high Signal-to-Noise Ratio (SNR) scenario, but very likely non line-of-sight (NLOS) due to the airport infrastructure. The other case is the registration at the cell border which conforms to a low SNR but line-of-sight (LOS) scenario. For the detection of RA frames within the RA channel, different detection methods are compared with each other via simulations. Furthermore, the necessary accuracy of the Doppler estimation to achieve a certain Bit Error Rate (BER) is determined and possible improvements of the RA frame are discussed.

Keywords: Aeronautical communication system · Synchronization · Doppler · LDACS · Random Access

1 Introduction

Due to increasing numbers of aircraft [2], nowadays aeronautical communication systems will come to their limits in the near future. The situation becomes even worse since it is planned to reduce flight time by leaving airways and taking more direct routes, requiring adjusting flight plans in real-time as well as shortening the separation distance between aircraft [3]. All together, this results in an additionally increased communication. That is the reason why new communication systems like LDACS become necessary. One crucial part for the communication setup is the registration of an AS to the system itself via the RA channel, as it takes quite long if it is not possible for the AS to register reliably.

The outline of this paper is as follows. In Sect. 2 a brief introduction of the LDACS system design, the implemented simulator and the different flight scenarios is given, as well as the examined RA frame detection methods are shown. Simulation results for RA frame detection and the impact onto the BER due to a wrong Doppler shift estimation are presented in Sect. 3. An improvement for the RA frame for a more reliable detection is given in Sect. 4.

M. Ma (Ed.): ICTCE 2020, LNEE 797, pp. 26–34, 2022.
https://doi.org/10.1007/978-981-16-5692-7_4

2 LDACS System Design and Simulation Environment

LDACS is an orthogonal frequency-division multiplexing (OFDM) based terrestrial communication system with an effective channel bandwidth of 500 kHz, consisting of an up- and down-link called Forward Link (FL) and Reverse Link (RL). Both links are frame-based and have the so-called Super-Frame (SF) with a periodicity of 240 ms in common (see Fig. 1). All GS are synchronized with each other, and therefore build the common time reference for the RL.

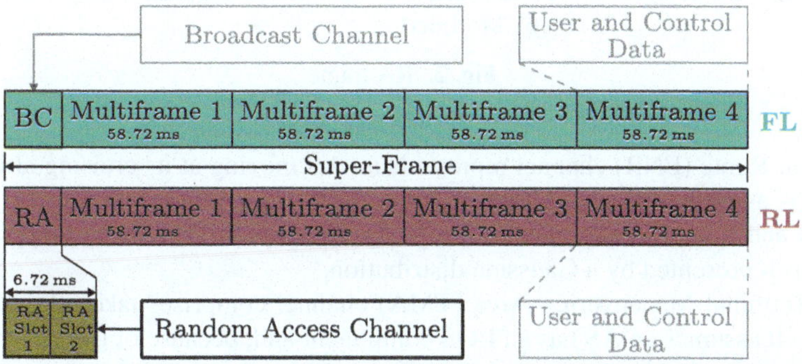

Fig. 1. FL and RL framing structure

2.1 Random Access Procedure

When an unregistered AS has found the start of the SF, it sends an RA frame depicted in Fig. 2(a) to register itself to the system. For this purpose, it has two opportunities to send an RA frame, by selecting one of the two RA slots, the used procedure is slotted ALOHA [4]. The temporal length of an RA slot corresponds to the largest cell size of an LDACS GS with a cell radius of 200 NM. The propagation delay of the FL to the outermost AS and back plus the frame length of the RA frame of 840 μs is 3:31 ms, which fits within one RA slot.

The RA frame (see Fig. 2(a)) consists of an Automatic Gain Control (AGC) symbol which is necessary, since the receiver is normally set to highest sensitivity to detect an AS at the cell border. After the AGC symbol, two synchronization symbols follow, a short and long symbol for time and frequency estimation. The remaining symbols hold the user data. The Peak to Average Power Ratio (PAPR) subcarriers are used to reduce the crest factor. The pilot subcarriers and the synchronization symbols are necessary for channel estimation, recommendable over the two-dimensional time/frequency grid.

Table 2 gives an overview of the used coding parameter and modulation scheme for the standardized RA frame.

2.2 Flight Scenarios/Channel Types

In the LDACS standard, three different flight scenarios for the L-Band have been defined [5], corresponding to three different channel types.

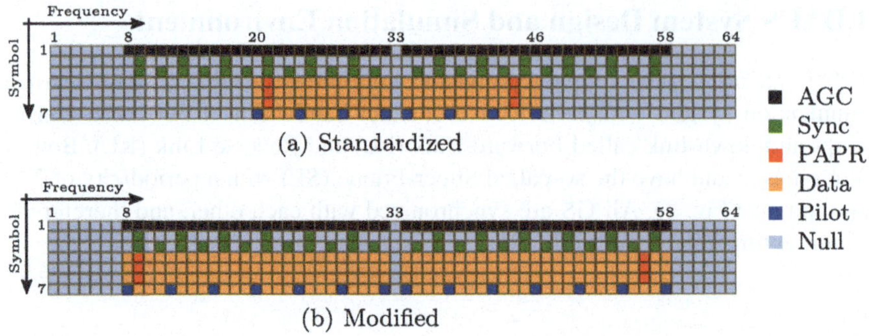

Fig. 2. RA frame

The En-Route (ENR) channel represents an AS moving at its cruising altitude and speed. It is assumed that the AS has a strong LOS component, followed by a nearby reflection and a large-scale reflection, e.g., by a mountain. The mean Doppler of the scatterer is represented by a Gaussian distribution.

The Terminal Maneuvering Area (TMA) channel comprises take-off and landing scenarios. It assumes an AS has an LOS component and, because of the lower altitude, many uniformly distributed scatterer. The Power Delay Pro le (PDP) of the scatterer is modeled as an exponentially decaying Rician fading process. Due to the distribution of the scatterer the Doppler spectrum is of Jakes type.

The Airport (APT) channel represents taxiing and parking of an AS at the airport and, thus, has quite often no LOS component because of buildings and other infrastructure on the ground. The PDP is modeled as an exponentially decaying Rayleigh fading process with strong multipath components. The scatterer are uniformly and uncorrelated distributed around the AS, which leads again to a Jakes type Doppler spectrum.

All defined channel parameters can be found in Table 1 [1, 5].

2.3 LDACS Simulator Environment

The proposed simulation environment, depicted in Fig. 3, is able to simulate FL, RL, and RA scenarios. The simulator is implemented in MATLAB, and can be split in three parts. The transmitter contains the entire LDACS coding and modulation scheme according to the LDACS standard [1]. The smallest simulation unit for FL and RL is a Super-Frame. For the RA, it is one RA slot.

The LDACS signal is passed through the channel as stated in Sect. 2.2, and additive white Gausian noise (AWGN) is added. The third part of the simulator is the receiver, where time and frequency synchronization is performed. Channel estimation is done over a two-dimensional OFDM time/frequency grid via an inverse weighting method [6]. Subsequently, the received signal is demodulated and decoded. The received data is compared with the sent one and the BER is determined. For the synchronization to an RA frame, it is assumed that the frame is at a random position within an RA slot. For this purpose, different synchronization methods are compared with each other (see Sect. 2.4).

Table 1. Channel parameters

Channel	Fading	Delay	Doppler
ENR	Rician: $k_\mathrm{R} = 15\,\mathrm{dB}$ Near-Specular to Off-path: $k_\mathrm{NS/OS} = 6\,\mathrm{dB}$	$\tau_0 = 0.3\,\mu\mathrm{s}$ $\tau_1 = 15\,\mu\mathrm{s}$	$f_{\mathrm{D_{LOS}}} = 1700\,\mathrm{Hz}$ Mean Doppler Spec. $f_{\mathrm{M_0}} = 0.85 \times f_{\mathrm{D_{LOS}}}$ $f_{\mathrm{M_1}} = -0.6 \times f_{\mathrm{D_{LOS}}}$ Doppler Spread Spec. $f_{\mathrm{S_0}} = 0.05 \times f_{\mathrm{D_{LOS}}}$ $f_{\mathrm{S_1}} = 0.15 \times f_{\mathrm{D_{LOS}}}$
TMA	Rician: $k_\mathrm{R} = 10\,\mathrm{dB}$	exp. decay $\tau_\mathrm{max} = 10.5\,\mu\mathrm{s}$	$f_{\mathrm{D_{LOS}}} = 624\,\mathrm{Hz}$ Jakes Spectrum for Scatterer
APT	Rayleigh: $k_\mathrm{R} = -100\,\mathrm{dB}$	exp. decay $\tau_\mathrm{max} = 3\,\mu\mathrm{s}$	$f_\mathrm{D} = 413\,\mathrm{Hz}$ Jakes Spectrum

2.4 Synchronization Methods

A lot of OFDM synchronization methods have been published over the last years. Some of them are very simple, which keeps the computational complexity low. Other algorithms are more suited for low SNR scenarios but have, in contrast, a pretty high computational complexity. In this section, several synchronization methods for detecting an RA frame are compared with respect to computational complexity and performance.

Schmidl & Cox (SC) [7]. This popular low complexity algorithm is based on repeating time domain sequences of an OFDM signal, the so-called synchronization symbols. The start of the symbol is found when the repeating parts correlate with each other. An additional advantage of this algorithm is that the phase of the correlation can be used for a rough Doppler shift estimation. A minor drawback is that the correlation results in a plateau in the length of the cyclic prefix and not in a distinct peak. Hence, fine synchronization via the synchronization symbols is necessary afterwards. The computational complexity for the correlation equals 8 real multiplications plus 8 real additions per time sample, independent on the oversampling factor.

Abdzadeh-Ziabari et al. Method [8]. This algorithm is based one Kang's method [9] but has the advantage of extended correlation of the synchronization symbol. Kang's method uses a so-called "correlation sequence of a preamble" [9], formed by cyclically shifting the synchronization symbol in time domain. The shifted version with the "most impulsive" [9] auto-correlation is used to get a distinct correlation peak. Both the transmitter and the receiver side have to know the selected shift. Abdzadeh-Ziabari et al.'s method as the extension of Kang's method uses more than one shifted versions of the synchronization symbol. By taking multiple time shifted versions, the correlation length is increased and one gets a more trustworthy correlation peak. A general equation for the computational complexity is not possible. Abdzadeh-Ziabari et al. states the "Reduced Complexity Estimator" [8] which gives the computational complexity dependent on the number of used sequences $\mathrm{l_s}$ and the correlation length L. L is in the range of $N \leq L$

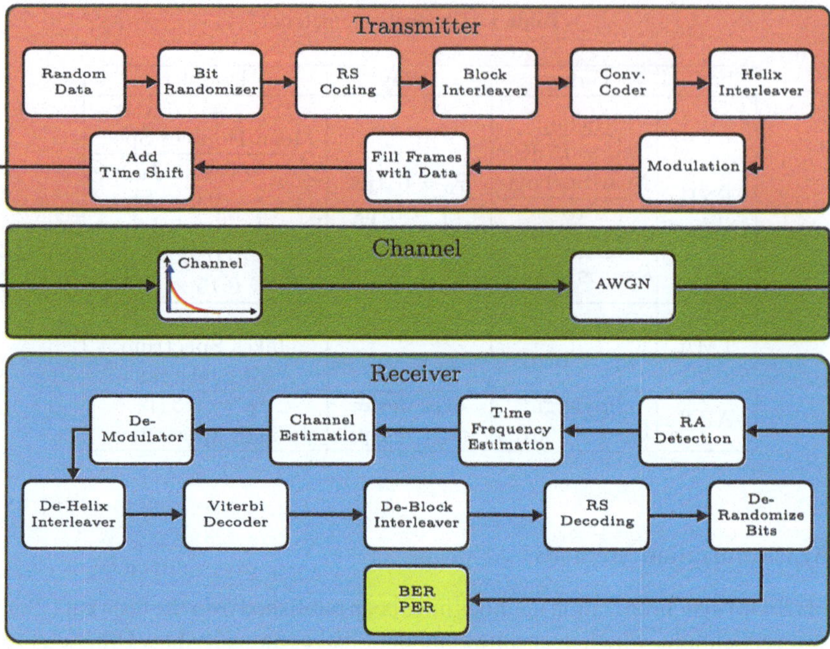

Fig. 3. Block diagram of LDACS system simulator

$\leq (N-1) \times N/2)$, where N is the symbol length. $\mathrm{l_s}$ depends on the correlation length L, and is in the range of $2 \leq l_s \leq (N-1)$. This results in $4 \times l_s$ real multiplications and $2 \times (L + l_s - 1)$ real additions per sample. For the conducted simulations with $L = 512$ and $l_s = 9$, its 36 multiplications and 1040 additions.

Matched Filter (MF). This commonly used method for detecting known signals gives a distinct correlation peak if the received signal and MF correlate in time domain. One drawback of the MF is that the correlation peak degenerates quite fast if the Doppler shift is too high. A possible countermeasure is to use multiple MFs, shifted by certain frequency o sets to detected the Doppler shifted versions of the signal. Furthermore, the multipath path components of the channel will result in multiple detection peaks. To enhance the received SNR, a rake receiver could be used. Another drawback is that the computational complexity is rather high with 256 real multiplications plus 254 real additions per time sample for an oversampling factor of one. In combination with possibly Doppler shifted versions of the MF, a practical implementation could be by far too complex. For the simulations in Sect. 3.1, the MF is used as reference design for RA frame detection, but without the use of multiple Doppler shifted MF versions.

3 Simulation Results

This section presents the results of the conducted simulations. The first subsection concerns the reliable detection of an RA frame. The second subsection presents the necessary accuracy of the Doppler shift estimation for reliably decoding an RA frame.

3.1 Random Access Synchronization

For the simulations, an RA frame is assumed to be somewhere randomly distributed within one RA slot. All detection methods presented in Sect. 2.4 are applied to the same received RA frame. The collision of multiple frames within an RA slot and the resulting impact on the detection algorithms is not simulated. A synchronization counts as successful, if the method is able to detect the start of a frame within the Cyclic Prefix (CP) of the synchronization symbol, otherwise it is counted as fail.

ENR Scenario. The ENR channel has, due to the cruising speed of the AS, the highest Doppler shift (see Table 1). The synchronization results are depicted in Fig. 4(a), and they show that the SC method has the lowest probability performance to correctly detect the start of an RA frame. Abdzadeh-Ziabari et al.'s method works quite well for the ENR case and low SNR. Surprisingly, the MF has the best detection results, despite the fact of a high Doppler shift. The simulation result of the SC is not that worse as it looks like, because the LDACS standard states a minimum SNR for FL operations of $\approx$3 dB to a achieve a maximum BER of 10^{-6}. For RL operations, even an SNR of $\approx$5 dB is necessary. Thus, for the SC method more than 90% of the RA frames will be detected.

TMA Scenario. Synchronization during take-off/landing is quite unusual, since the AS should register at the airport or at the cell-border. For the sake of completeness this scenario is also simulated. Figure 4(b) shows that all methods work quite well for the TMA scenario. One has to keep in mind that the SNR for this scenario is already quite high.

APT Scenario. This channel type represents the worst case synchronization scenario for LDACS. As depicted in Fig. 4(c), all methods need a high SNR to achieve 100% detection probability. Nevertheless, one can assume a high SNR since the AS is close to the GS receiver. Again, the MF gives the best detection result, followed by the Abdzadeh-Ziabari et al.'s method.

Summary. All presented methods work well for the presented channel models. If a low computational complexity algorithm is desired, then SC is the method of choice. The performance for the ENR channel is a little bit lower compared to the other two methods. Abdzadeh-Ziabari et al.'s method works well but has a high computational complexity. For all scenarios the MF has the best performance. Although the computational complexity is very high, one should keep in mind that nowadays Field Programmable Gate Array (FPGA) are quite powerful. Furthermore, only one MF was used for simulations, without additional Doppler shifted versions.

3.2 Necessary Doppler Accuracy

The goal of this simulation is to determine the BER performance of the RA frame, with respect to accuracy of the Doppler shift estimation via the two synchronization symbols. It is assumed that the start of an RA frame is already known to the receiver. The Doppler shift for the corresponding channel, according to Table 1 is used. For this simulation,

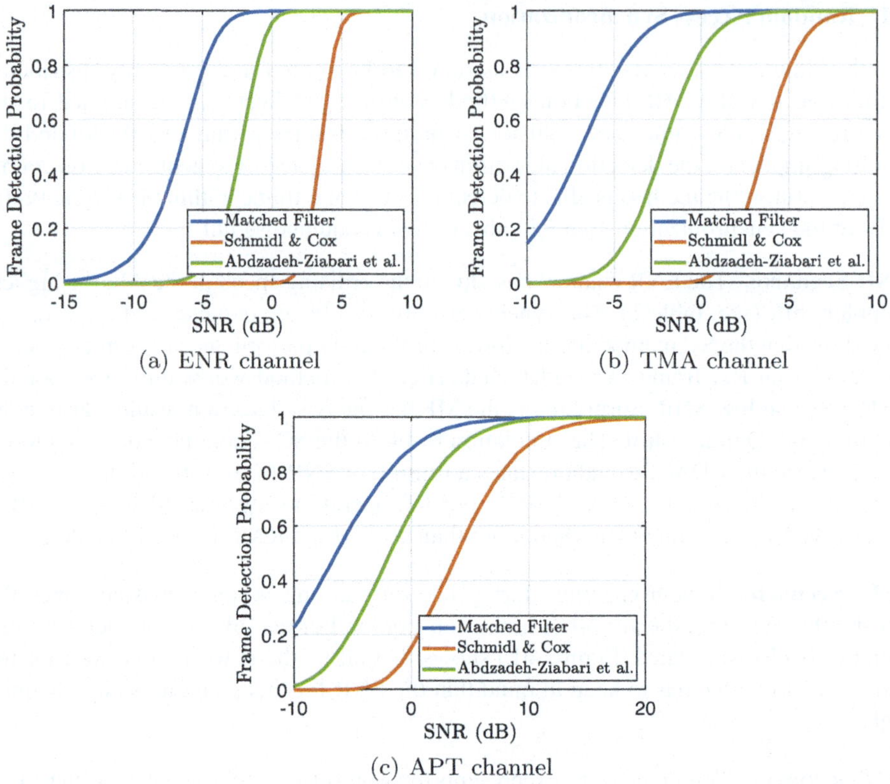

Fig. 4. RA frame detection

the Doppler shift is not estimated by the receiver side itself, the "estimated" Doppler shift is given by the simulator to the receiver for synchronization. This given "estimated" Doppler shift is varied relative to the true Doppler shift. Figure 5(a) shows the necessary Doppler estimation accuracy for the three defined channel types (see Sect. 2.2). For the ENR scenario, the estimation of the Doppler has to be at least as good as 400 Hz without a severe degeneration of the BER. For TMA channel, the simulation results show that it is not possible to stay under a certain BER, even if the Doppler shift is exactly known. The unsymmetrical shapes of the simulation results for ENR and TMA channel are due to the Doppler spread of the scattering components which differ from the LOS component.

Further investigation has shown that the results do not depend on the SNR. Even if it would be free of noise, simulation results are not getting better. As a conclusion, the channel models have a severe impact onto the BER of an RA frame. In regard of the APT channel, simulation results are even worse.

4 Possible RA Frame Improvements

Simulations of Sect. 3.2 have shown that with the standardized RA frame type (see Fig. 2(a)) it is not possible to detect the frame under a certain BER for some channel

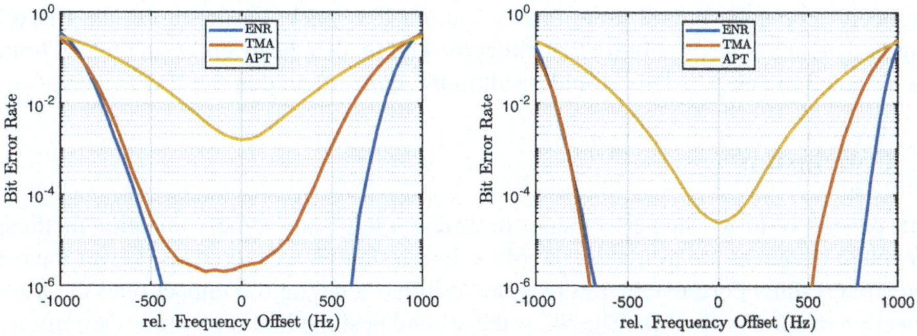

(a) Standardized RA Frame (b) Modified RA Frame

Fig. 5. BER vs. Doppler accuracy

types. One problem for the worse BER is that the coding is not as strong as it is in the RL/FL data transmissions, where several data packets are jointly coded. This results in a higher robustness against channels with strong multipath components. Since the RA has several unused subcarriers, it is possible to use them for a more robust channel coding scheme without adding an additional OFDM symbol and unnecessarily lengthening the RA frame. Figure 2(b) shows a modified RA frame with a 1/6 convolutional code rate instead of the 1/3 rate. Table 2 lists the parameters for the proposed RA frame modification. The parameter for the helical interleaver have been slightly changed, since more coded bits need to be interleaved. The listed values give a good spreading of the coded data bits over the time/frequency plane of the RA frame.

Table 2. Parameters for modified RA frame

	Modulation	Convol. coding rate	Number uncoded bits	Helical interleaver (a,b)	Number coded bits
Standard	QPSK	1/3	54	(15,12)	180
Modified	QPSK	1/6	54	(20,18)	360

Simulation results with the modified RA frame can be seen in Fig. 5(b). The necessary Doppler estimation accuracy for the ENR channel is relaxed to 700 Hz. For the TMA channel, the Doppler estimation accuracy has to be better than 500 Hz, to achieve a BER below 10^{-6}. For APT the results are better, but still do not achieve the target BER of 10^{-6}.

One drawback of the modified RA frame is the possible 3 dB increase of the PAPR. This could be a problem if the power amplifier for the AS will be a low cost solution and, therefore, the linearity for higher power is not given. This would introduce higher order intermodulation and broaden the output spectrum. With the PAPR carriers within the RA frame it is possible to reduce the occurring PAPR of the frame in advance. Since most of the data in the frame is fixed and, furthermore, the calculation of the PAPR does not need to be done in real-time, this is a minor drawback. Furthermore, the LDACS standard includes the possibility that either the half or all subcarriers of an OFDM frame are assigned to one AS. This would result in the same effect that the PAPR increases.

5 Conclusion

This paper has shown the performance of different RA frame synchronization methods. If enough resources are available, the MF solution should be favored, as it gives the best detection results. Fortunately, the Doppler shift has a negligible impact onto the frame detection results for the MF. The SC is the second best method if resources are limited. The algorithm is suitable for all defined channel models, although the performance for the ENR channel is not as good compared to Abdzadeh-Ziabari et al.'s method and the MF.

Furthermore, the necessary accuracy of the Doppler shift estimation and the impact on the BER was examined. It turned out that with the standardized RA frame the achievable BER is limited for the TMA and APT channel, even if the Doppler shift is exactly known to the receiver. Since the RA procedure is crucial to establish a connection to the communication system, it is necessary to reconsider the standardized RA structure. For this purpose, an improvement of the RA frame with a stronger channel coding has been presented in this paper, without changing the frame length. The simulation results show that the BER performance for the APT channel is limited for both RA frame types. One explanation for the results is the unrealistic high taxiing speed for the APT channel, which leads to a high Doppler shift (see Table 1).

References

1. Graupl, T., Rihacek, C., Haindl, B., Parrod, Q.: LDACS A/G specification. Technical report, D3.3.010, SESAR-JU, December 2018
2. Eurocontrol: European Aviation in 2040: Challenges of Growth, October 2018. https://www.eurocontrol.int/sites/default/les/2019-07/challenges-of-growth-2018-annex10.pdf
3. SESAR: SESAR project shows how aircraft separation methods can increase runway capacity. https://www.sesarju.eu/news/sesar-project-shows-how-aircraft-separation-methods-can-increase-runway-capacity
4. Abramson, N.: Packet switching with satellites. Technical report, Hawaii University, Honolulu (1973)
5. German Aerospace Center: D5 Deliverable: Expected B-AMC System Performance. Technical report, Eurocontrol, September 2007. https://www.eurocontrol.int/sites/default/les/2019-05/24092007-b-amc-project-deliverable-d5-v11.pdf
6. Shepard, D.: A two-dimensional interpolation function for irregularly-spaced data. In: Proceedings of the 1968 23rd ACM National Conference, ACM 1968, New York, NY, USA, pp. 517–524. ACM (1968). http://doi.acm.org/10.1145/800186.810616
7. Schmidl, T.M., Cox, D.C.: Robust frequency and timing synchronization for OFDM. IEEE Trans. Commun. **45**(12), 1613–1621 (1997)
8. Abdzadeh-Ziabari, H., Shayesteh, M.G., Manaffar, M.: An improved timing estimation method for OFDM systems. IEEE Trans. Consum. Electron. **56**(4), 2098–2105 (2010)
9. Kang, Y., Kim, S., Ahn, D., Lee, H.: Timing estimation for OFDM systems by using a correlation sequence of preamble. IEEE Trans. Consum. Electron. **54**(4), 1600–1608 (2008)

Energy Efficient Power Allocation in Massive MIMO Systems with Power Limited Users

Abdolrasoul Sakhaei Gharagezlou[1], Mahdi Nangir[1(✉)], Nima Imani[1], and Erfan Mirhosseini[2]

[1] Electrical and Computer Engineering Department, University of Tabriz, Tabriz, Iran
nangir@tabrizu.ac.ir

[2] Electrical Engineering Department, Amirkabir University of Technology, Tehran, Iran
erfanm@aut.ac.ir

Abstract. In this paper, we investigate energy-efficient (EE) power allocation (PA) for a special downlink scenario of the massive multiple-input multiple-output (MIMO) systems. We consider a minimum power required for each user to ensure that the quality of service (QoS) for each user is satisfied. In this method, a comparison between the sum of minimum power required by users and the maximum transmission power is done to determine whether maximizing EE is possible or not. If the sum of the minimum power required by users is less than the maximum transmission power, we maximize EE. Otherwise, the number of admitted users in a cluster is maximized. In both cases, the simulation results show that the proposed algorithm has better performance than similar algorithms.

Keywords: Massive multiple-input multiple-output · Energy efficiency · Power allocation · Power limited user · Admitted user

1 Introduction

The MIMO systems, in which a base station (BS) is equipped with hundreds of antennas and communicates with tens or hundreds of users simultaneously in a same frequency-time block, are known as massive MIMO systems. As the number of BS antennas in massive MIMO systems goes to infinite, the noise and dimming effects disappear in the small scale range. Massive MIMO systems provide reliable communication, high energy efficiency and low-complexity schemes for signal processing. Several technologies are involved in the massive MIMO in fifth generation (5G) networks including non-orthogonal multiple access (NOMA), heterogeneous networks, millimeter waves and device-to-device communications [1].

The total sum-rate obtained by the power allocation method in multi-cell massive MIMO systems is studied in [2]. A new power allocation algorithm is proposed to increase the energy efficiency and capacity of massive MIMO systems [3]. A two-step iterative algorithm with a combination of antennas and users is proposed to maximize energy efficiency in [4]. The results presented in this paper show that energy efficiency with the maximum-ratio combining (MRC) receiver has been improved by 71.16%

M. Ma (Ed.): ICTCE 2020, LNEE 797, pp. 35–46, 2022.
https://doi.org/10.1007/978-981-16-5692-7_5

when the number of users equals 60. In [5], the authors examine allocation of the energy efficient power for the massive MIMO system with maximum ratio transmission (MRT) pre-coding scheme [5]. The problem of specialized power optimization in radio cognitive networks using the NOMA method is investigated [6]. In [7], to solve the power allocation problem for a MIMO-NOMA system in a cluster, a minimum rate is considered for each user. An optimal power allocation solution that can improve the total data rate of a mobile network cell with a reduced complexity is proposed in [8]. In [9], the minimum transmission power of each user is calculated using the maximal signal to leakage ratio (maximal-SLR) solution for each user. Furthermore, a new radio access scheme is proposed which consists of the relay protocol and the NOMA method [10]. A power allocation scheme is obtained for a set of parallel channels while the transmitter has partial channel status information [11]. The energy efficiency of a heterogeneous cellular network with massive MIMOs and Small Cells (SCs) is discussed in [12]. Furthermore, the energy-efficient power allocation for a multi-user massive MIMO system in a single cell is investigated by using the SIF method in [13].

In this paper, we propose an energy efficient power allocation scheme in the massive MIMO scenario which is motivated by [13]. The novelty and innovation of this paper can be summarized as follows.

In this work, a minimum power required by users is considered to provide quality of user service. The Optimization problem in this paper is divided into two parts. They are maximizing energy efficiency and maximizing the number of admitted users in a cluster. First, we determine which of the optimization problems should be addressed by comparing the sum of the minimum power required by users to ensure the quality of service with the maximum transmit power. If sum of the minimum power required by users is less than the maximum power output, the energy efficiency is maximized. Otherwise, the number of accepted users in a cluster is maximized. We propose a new iterative algorithm for the power allocation and consider a required power for satisfying the QoS of users in a cluster. It is shown that the proposed algorithm performs better than other similar power allocation algorithms.

The following four parts of this article are discussed below. In Sect. 2, the model of the system studied is described. In Sect. 3, the optimization problem is formulated and we obtain the optimal power allocation scheme by solving it. In Sect. 4, simulation results and performance analysis are presented. Finally, we conclude paper in Sect. 5.

2 System Model

In this paper, we consider the system depicted in Fig. 1. In this system, the BS is equipped with M antennas and each user is equipped with single antenna.

Let consider G is the matrix of the flat fading channel between the BS and K users. G can be formulated as:

$$\mathbf{G} = \mathbf{H}\mathbf{D}^{1/2}. \tag{1}$$

where, $\mathbf{H} \epsilon C^{M \times K}$ is the small scale fading channel matrix and $\mathbf{D} = \mathrm{diag}\{\beta_1.\beta_2.\ldots.\beta_K\}$ denotes the large scale fading matrix, where $\beta_K = \varphi\varrho/d_k^{\varepsilon}$ represents the path loss and shadow fading. Parameter φ is a constant related to the carrier frequency and antenna gain and d_k is the distance between the BS and the k-th user. Furthermore, ε represents the path loss exponent, and ϱ is the shadow fading with lognormal distribution, i.e., $10\log_{10}(\varrho) \sim N(0.\sigma^2)$.

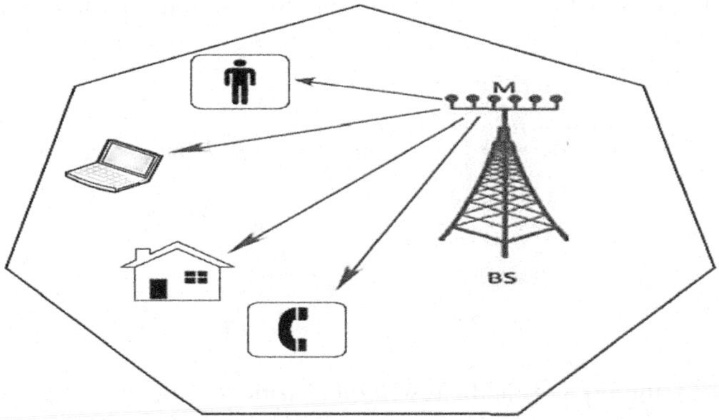

Fig. 1. System model of the massive MIMO systems.

Accordingly, the observed signal of the k-th user is formulated as follows [13]:

$$y_k = \sqrt{p_k\beta_k}\|\boldsymbol{h}_k\|s_k + \sum_{l=1.l\neq k}^{K}\sqrt{p_l\beta_k}\frac{\boldsymbol{h}_k\boldsymbol{h}_l^H}{\|\boldsymbol{h}_l\|} + n_k. \quad (2)$$

where $k \in \{1.2.\ldots.K\}$, p_k denotes the transmit power allocated to the k-th user and $\boldsymbol{h}_k$ is the k-th column of $\mathbf{H}$. Besides, s_k represents transmit data symbol of the k-th user and n_k is the Additive White Gaussian Noise (AWGN) at the k-th user with distribution $N(0.N_0)$, where N_0 denotes the noise power spectral density.

The received SINR for k-th user can be expressed as:

$$\gamma_k = \frac{p_k\beta_k\|\boldsymbol{h}_k\|^2}{\left\|\boldsymbol{h}_k\sum_{l=1.\neq k}^{K}\sqrt{p_l\beta_k}\frac{\boldsymbol{h}_l^H}{\|\boldsymbol{h}_l\|}\right\|^2 + \sigma^2}. \quad (3)$$

where $\|.\|$ represents the $L2$-norm. The spectral efficiency for k-th user can be written as:

$$r_k = B\log_2\left(1 + \frac{\gamma_k}{\mu}\right). \quad (4)$$

where $\mu = \frac{2}{3}\ln(5P_e)$ is the SNR gap between the Shannon channel capacity and a practical modulation with coding scheme which achieves the BER value of P_e [14]. Then, the total spectral efficiency achieved by all users can be expressed as:

$$r(\boldsymbol{P}) = \sum_{k=1}^{K} r_k = B\sum_{k=1}^{K}\log_2(1 + \frac{\gamma_k}{\mu}). \quad (5)$$

where $\boldsymbol{P} = [p_1.p_2.\ldots.p_K]^T$ is the power allocation vector for all users from the BS.

According to description of (1), $\boldsymbol{h}_l$ $(l = 1.2.\ldots.k-1.k+1.\ldots.K)$ are independent of random vector $\boldsymbol{h}_k$. If we define $\alpha_l \cong \left\| \frac{\boldsymbol{h}_k \boldsymbol{h}_l^H}{\|\boldsymbol{h}_l\|} \right\|^2$, then α_l will be a gamma random variable with parameters (1,1) and $\alpha_0 \cong \|\boldsymbol{h}_k\|^2$ is a gamma random variable with parameters $(M,1)$. Thus we have,

$$E\left[\left\| \boldsymbol{h}_k \sum_{l=1.\neq k}^{K} \sqrt{p_l \beta_k} \frac{\boldsymbol{h}_l^H}{\|\boldsymbol{h}_l\|} \right\|^2\right] = \beta_k \sum_{l=1.\neq k}^{K} \sqrt{p_l}\, E[\alpha_l] = \beta_k \sum_{l=1.\neq k}^{K} \sqrt{p_l}. \tag{6}$$

and

$$E\left[p_k \beta_k \|\boldsymbol{h}_k\|^2\right] = M p_k \beta_k. \tag{7}$$

Using (6) and (7), the Eq. (3) can be rewritten as follows [5]:

$$\gamma_k = \frac{p_k \beta_k M}{\beta_k \sum_{l=1.\neq k}^{K} \sqrt{p_l} + \sigma^2}. \tag{8}$$

In this work, the energy efficiency is defined as follows:

$$\eta_{EE} = \frac{\sum_{k=1}^{K} r_k}{\sum_{k=1}^{K} p_k + \sum_{m=1}^{M} P_{c.m}}. \tag{9}$$

where $P_{c.m}$ is the fixed circuit power consumption per antenna.

Our goal is to maximize the energy efficiency when each user has a pre-defined minimum required power. The optimization problem is formulated as follows:

$$maximize_{\{p_1.p_2.\ldots.p_K\}} \ \eta_{EE} \tag{10.a}$$

$$C1\colon \sum_{k=1}^{K} p_k = P_T \tag{10.b}$$

$$C2\colon p_k \geq (\omega_k - 1)\left(\sum_{j=1}^{k-1} p_j + \frac{1}{\vartheta \|\boldsymbol{h}_k\|^2}\right). \tag{10.c}$$

where (10.b) shows the total BS power constraint and P_T denotes the flexible transmit power. The constraints of (10.c) show QoS of users, i.e., the minimum required power, is guaranteed. In C_2, $\omega_k = 2^{R_k^{\text{Min}}}$ where R_k^{Min} is the minimum required data rate for k-th user for $k = 1.2.\ldots.K$. Furthermore, ϑ denotes the signal-to-noise ratio (SNR).

3 Proposed Solution

Given the minimum power for each user (10.c), the optimization problem (10) may not a solution if the total transmission power is not large enough. If the problem of maximizing energy efficiency is not feasible, then we maximize the number of users in the cluster and hereto we provide an efficiency.

We compare sum of the minimum power required by users with the total transmission power for determining the feasibility of the optimization problem (10). The minimum power required by each user is equal to:

$$P_{req.k} = (\omega_k - 1)\left(\sum_{j=1}^{k-1} p_j + \frac{1}{\vartheta \|\boldsymbol{h}_k\|^2}\right). \tag{11}$$

which is the minimum required power to satisfy the QoS requirement the of k-th user. As a result, if

$$\sum_{k=1}^{K} P_{req.k} \le P_{max}. \tag{12}$$

then the problem (10) is feasible; otherwise, it is not.

3.1 EE Maximization When Problem (10) is Feasible

The problem of maximizing energy efficiency can be formulated as:

$$maximize_{\{p_1.p_2.....p_K\}} \quad \eta = \frac{R^{sum}}{P_c + \sum_{k=1}^{K} p_k} \tag{13.a}$$

$$C2: \sum_{k=1}^{K} p_k = P_T \tag{13.b}$$

$$C3: p_k \ge (\omega_k - 1)\left(\sum_{j=1}^{k-1} p_j + \frac{1}{\vartheta \|\boldsymbol{h}_k\|^2}\right). \tag{13.c}$$

The objective function of (13) has a fraction form and it is a non-convex function. Using fractional programming and low data rate, problem (13) can be turned into a convex optimization problem. The maximum EE can be achieved if and only if,

$$maximize_{\{p_1.p_2.....p_K\}} \left\{\sum_{k=1}^{K} r_k - q^*\left(\sum_{k=1}^{K} p_k + \sum_{m=1}^{M} P_{c.m}\right)\right\} = 0. \tag{14}$$

where $q*$ represents the maximum EE [15]. By considering perfect channel state information (CSI), the Rayleigh fading and the MRT pre-coding, the lower bound of the data rate can be written as follows [5],

$$\tilde{r}_k = B\log_2\left(\frac{M\beta_k p_k}{\beta_k \sum_{\substack{j=1 \\ j \neq k}}^{K} p_j + \sigma^2}\right). \tag{15}$$

According to (15), the optimization problem in (13) can be converted to a simpler form as:

$$maximize_{\{p_1.p_2.\ldots.p_K\}}\left\{\sum_{k=1}^{K}\tilde{r}_k - q^*\left(\sum_{k=1}^{K}p_k + \sum_{m=1}^{M}P_{c.m}\right)\right\} \tag{16.a}$$

$$C1{:}\ \sum_{k=1}^{K}p_k = P_T \tag{16.b}$$

$$C2{:}\ p_k \geq (\boldsymbol{\omega}_k - 1)\left(\sum_{j=1}^{k-1}p_j + \frac{1}{\vartheta\,\|\boldsymbol{h}_k\|^2}\right). \tag{16.c}$$

The simplified optimization problem in (16) is a constrained problem, hence the Lagrange function is used to convert it to a non-constrained problem [16]. Let,

$$\begin{aligned}\Phi(\mathbf{p}.\theta.\lambda) = &-\left[\sum_{k=1}^{K}r_k - q\left(\sum_{k=1}^{K}p_k + \sum_{m=1}^{M}P_{c.m}\right)\right]\\ &-\lambda_k\left(p_k - (\boldsymbol{\omega}_k - 1)\left(\sum_{j=1}^{k-1}p_j + \frac{1}{\vartheta\,\|\mathbf{h}_k\|^2}\right)\right)\\ &-\theta\left(P_T - \sum_{k=1}^{K}p_k\right)\end{aligned} \tag{17}$$

Necessary and sufficient condition to obtain the optimal transmission power is expressed as:

$$\begin{aligned}\frac{\partial\phi}{\partial p_k} = &\sum_{j=1.j\neq k}\frac{1}{\left(\sum_{.i=1\neq k}^{K}p_i + \sigma^2/\beta_i\right)ln2} - \frac{1}{p_k ln2}\\ &+\left(\boldsymbol{\lambda}_k - \sum_{j=1}^{k-1}(\omega_j - 1)\boldsymbol{\lambda}_j\right) + \theta + q = 0\end{aligned} \tag{18}$$

Thus by using (18), the optimal power of each user is formulated as:

$$p_k = \frac{1}{\left(\sum_{j=1.j\neq k}\frac{1}{\left(\sum_{.i=1\neq k}^{K}p_i + \sigma^2/\beta_i\right)ln2} + \theta + q + \chi\right)ln2}. \tag{19}$$

where

$$\chi = \left(\boldsymbol{\lambda}_k - \sum_{j=1}^{k-1} (\omega_j - 1) \boldsymbol{\lambda}_j \right). \tag{20}$$

3.2 User Admission When Problem (10) is Infeasible

If (10) is infeasible, then the problem of maximizing admitted users is interested and it is formulated as follows:

$$maximize_{\{p_1 \cdot p_2 \cdot \ldots \cdot p_K\}} \sum\nolimits_{k=1}^{K} X_k \tag{21.a}$$

$$C1: \sum_{k=1}^{K} p_k = P_T \tag{21.b}$$

$$C2: p_k \geq (\boldsymbol{\omega}_k - 1) \left(\sum_{j=1}^{k-1} p_j + \frac{1}{\vartheta \left\| \boldsymbol{h}_k \right\|^2} \right) \tag{21.c}$$

$$C3 : X_k \in \{0.1\}. \tag{21.d}$$

where X_k is the binary decision variable indicating whether k-th user is admitted or not.

The user admission method is performed according to an iterative action. During each iteration, the user with the best channel gain is first selected. Among the selected users, their required power is calculated based on the interference of other accepted users. In the next step, this obtained power is compared with the total remaining power. If the total remaining power is higher than it, then this user is selected for acceptance in the cluster and it is removed from the candidates. Eventually, the remaining total power is updated. This process continues until no user meets the acceptance requirements for admission in the cluster [7]. The total remaining power is formulated as follows:

$$P_{remain} = P_{max} - P_{req}. \tag{22}$$

4 Simulation Results

In this section, simulation results are presented to verify the performance of the proposed PA strategy and user admission scheme. The parameters used in our simulations are given in Table 1.

In the following, we provide our proposed algorithm in a step by step procedure.

Table 1. Simulation parameters

Parameters	Font size and style
RB bandwidth B	120 kHz
Number of transmit antennas M	128
Number of users K	3
Noise spectral density N_0	−170 dBm/Hz
Variance of log-normal shadow fading σ^2	10 dB
Factor φ	1

Algorithm 1 Energy efficient power allocation algorithm for the massive MIMO system.

Initialize 1: $R_k^{Min}.(k = 1.2.....K)$

If_1 $(\sum_{k=1}^{K} p_{k.min} \le P_{max})$

Step 1: Initialize 2: $u = 0$, $v = \frac{1}{\mu\sigma^2 \ln 2}\underbrace{min}_{k}\{\beta_k\|\mathbf{h}_k\|^2\}$, $\varepsilon > 0$ $p_k > 0$ $(k = 1.2 ....K)$ transmit power and Lagrangian multipliers P_T, $\lambda^{(0)}$, $\theta^{(0)}$

Step 2: Let $\eta_{EE} = \frac{u+v}{2}$, solve p_k according to formula (19),

Then a realistic EE **q** can be got by formula (6),

If_2 $(\eta_{EE} \ge \eta_{EE}^{R})$, then $u = \eta_{EE}$

Else $v = \eta_{EE}$.

End if_2

Step 3: Update

$$\theta^{(n+1)} = max\left(0.\theta^{(n)} - \varpi_1\left(P_T - \sum_{k=1}^{K} p_k^{(n)}\right)\right)$$

$$\lambda^{(n+1)} = \max(0.\lambda^{(n)} - \varpi_k(p_k - (\omega_k - 1)\left(\sum_{j=1.j\ne k}^{K} p_j + \frac{1}{\vartheta\|h_k\|^2}\right))$$

Step 4: if_3 $(v - u < \varepsilon)$ then q $\approx \frac{u+v}{2}$

Else go to Step 2.

End if_3

n= n+1

End if_1

If_4 $(\sum_{k=1}^{K} p_{k.min} > P_{max})$

Step 1: User admission

End if_4

In this algorithm, two scenarios are considered. If maximizing the energy efficiency is not possible, the user conditions are considered in that user acceptance scenario in the cluster, unlike previous studies. Thus, a user which does not meet the minimum power requirements will not be left out, and generally the proposed algorithm outperforms other algorithms.

In Fig. 2, we compare the performance of the proposed algorithm with the proposed algorithm of [13]. As it is seen in Fig. 2, the proposed algorithm performs better than the SIF method. In this comparison, it is assumed that the total transmission power is 1 W. For instance, the energy efficiency of the proposed method is 6.84 Mbit/j for a fixed power value of 4 dBm, which has been improved about 1.23 Mbit/j compared to the proposed method of [13].

In Fig. 3, we present how the number of antennas affects the value of EE. In this implementation, it is assumed that the constant power of the circuit is 7 dBm. Obviously, in Fig. 3, the proposed method outperforms the method which has been proposed in [13]. For instance, the energy efficiency of the proposed method is 4.75 Mbit/j for the case of 64 antennas, which has been improved about 0.99 Mbit/j compared to the proposed method of [13].

If the condition that the sum of the minimum required power of users is less than the total transmitted power is not met, we maximize the number of accepted users in the cluster.

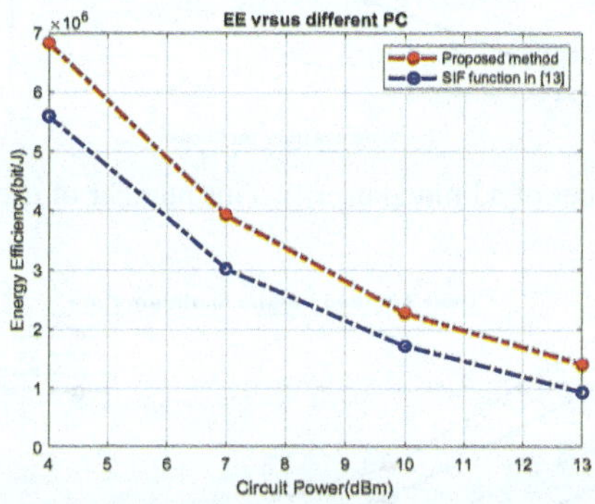

Fig. 2. Energy efficiency versus different P_C.

Figure 4 shows the average number of admitted users in a cluster. The number of users requesting to be added to the cluster is 9. As it is seen, this figure is plotted for 4 different scenarios, which shows that the average number of users accepted in the cluster increases with the maximum transmission power and the number of requesting users.

Figure 5 show the performance of the proposed user admission scheme versus minimum rate. Clearly, the average number of users accepted in a cluster decreases with increasing the minimum rate value.

Finally in Table 2, we compare the Energy Efficiency versus the maximum transmission power. According to the results appeared in this table, the proposed algorithm performs better than the SIF method which is proposed in [13].

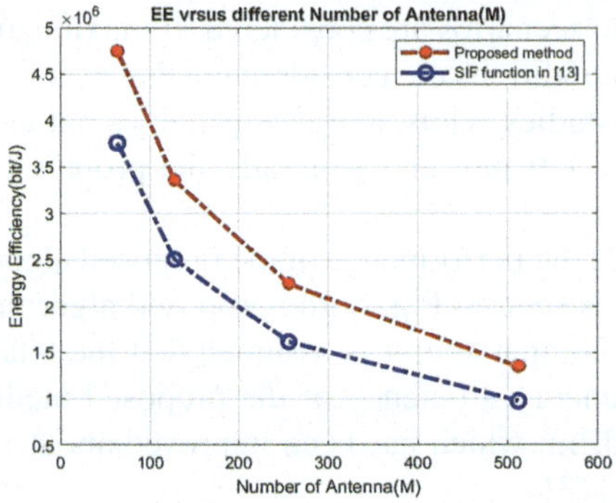

Fig. 3. Energy efficiency versus different M.

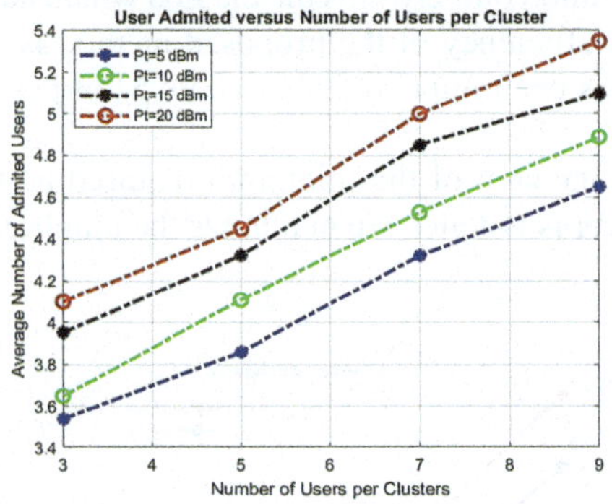

Fig. 4. Number of admitted users versus number of users in cluster.

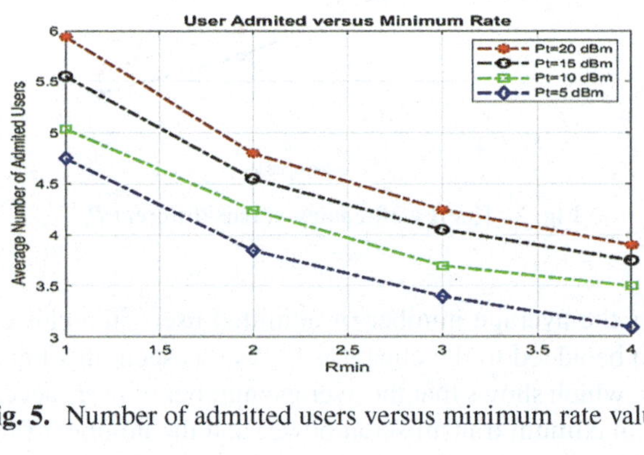

Fig. 5. Number of admitted users versus minimum rate value.

Table 2. Energy efficiency versus for different values of P_T.

Maximum transmission power (w)	EE convergence propose algorithm (Mbit/J)	EE convergence algorithm in [13] (Mbit/J)	Amount of improvement (Mbit/J)
2	4.6251	4.2127	0.4124
3	4.6027	4.2096	0.3931
4	4.5869	4.2021	0.3848

5 Conclusion

In this paper, we formulated the energy efficient power allocation problem for the massive MIMO system. In addition to the total transmission power, we also considered the minimum power required by users to ensure the quality of service for each user. By comparing the sum of the minimum power required by users and the total transmitted power, the type of optimization problem was determined. According to the presented results, we show that the proposed algorithm performs better than other similar power allocation algorithms.

References

1. Jameel, F., Haider, M.A.A., Butt, A.A.: Massive MIMO: a survey of recent advances, research issues and future directions. In: International Symposium on Recent Advances in Electrical Engineering (RAEE). IEEE (2017)
2. Zhang, Q., et al.: Power allocation schemes for multicell massive MIMO systems. IEEE Trans. Wirel. Commun. **14**(11), 5941–5955 (2015)
3. Honggui, D., Jinli, Y., Gang, L.: Enhanced energy efficient power allocation algorithm for massive MIMO systems. In: 11th International Conference on Communication Software and Networks (ICCSN). IEEE (2019). http://www.springer.com/lncs. Accessed 21 Nov 2016
4. Guo, M., Gursoy, M.C.: Energy-efficient joint antenna and user selection in single-cell massive MIMO systems. In: Global Conference on Signal and Information Processing (GlobalSIP). IEEE (2018)
5. Zhao, L., et al.: Energy efficient power allocation algorithm for downlink massive MIMO with MRT precoding. In: 78th Vehicular Technology Conference (VTC Fall). IEEE (2013)
6. Zeng, M., et al.: Power allocation for cognitive radio networks employing non-orthogonal multiple access. In: Global Communications Conference (GLOBECOM). IEEE (2016)
7. Zeng, M., et al.: Energy-efficient power allocation for MIMO-NOMA with multiple users in a cluster. IEEE Access **6**, 5170–5181 (2018)
8. Fan, L., et al.: Power control and low-complexity receiver for uplink massive MIMO systems. In: CIC International Conference on Communications in China (ICCC). IEEE (2014)
9. Fadhil, M., et al.: Power allocation in cooperative NOMA MU-MIMO beamforming based on maximal SLR precoding for 5G. J. Commun. **14**(8), 676–683 (2019)
10. Do, D.-T., Nguyen, T.-T.: Fixed power allocation for outage performance analysis on AF-assisted cooperative NOMA. J. Commun. **14**(7), 560–565 (2019)
11. Razoumov, L., Miller, R.R.: Power allocation under transmitter channel uncertainty and QoS constraints: volumetric water-filling solution. J. Commun. **7**(9), 656–659 (2012)

12. He, L., et al.: Maximizing energy efficiency in heterogeneous cellular network with massive MIMO and small cells. J. Commun. **11**(7), 616–623 (2016)
13. Zhang, J., et al.: Energy efficient power allocation in massive MIMO systems based on standard interference function. In: 83rd Vehicular Technology Conference (VTC Spring). IEEE (2016)
14. He, C., et al.: Energy efficiency and spectral efficiency tradeoff in downlink distributed antenna systems. IEEE Wirel. Commun. Lett. **1**(3), 153–156 (2012)
15. Dinkelbach, W.: On nonlinear fractional programming. Manage. Sci. **13**, 492–498 (1967)
16. Boyd, S.P., Vandenberghe, L.: Convex Optimization. Cambridge University Press, Cambridge (2009)

A High-Gain Low-Noise RF Front-End Design for WLAN Receivers

Haoran Xiong[1(✉)] and Kehan Huang[2]

[1] Southwest Jiaotong University, Chengdu 610000, China
2017114187@my.swjtu.edu.cn
[2] Australian National University, Canberra ACT 2600, Australia

Abstract. This paper proposes a novel design of a high power gain, low noise WLAN receiver. We designed the front-end of the receiver and the system can down-convert a typical WLAN signal to an IF frequency of 500 MHz. The proposed design has a gain of more than 54 dB and a noise figure of less than 8 dB. The RF front end for the WLAN receiver features an IIP3 of higher than −29 dB, a sensitivity of −123.0 dB, and a dynamic range is 43.8 dB. The proposed design features a direct-conversion scheme. In this paper, a brief introduction of the on-going WLAN standard is discussed; the block designs and the complete system design are demonstrated with the targeted specifications and simulation results; and the difficulties and solutions of the design process are showed.

Keywords: High Gain · Receiver · RF front-end · WLAN

1 Introduction

There is a growing need in the world for high-performance wireless technologies and apparatus [1–5]. Receivers are one of the essential parts of wireless communication systems. Currently there are many different commercial wireless communication standards that have use in a wide range of applications. The WLAN standard is one of the commonly accepted ones, whose main use is for Wi-Fi. Using the WLAN standards, a wireless receiver can be used to access the internet through the wireless transfer of data [6].

The WLAN receivers are required to be compatible with IEEE standards such as 802.11b, 802.11n, and 802.11ac. For example, the 802.11b standard is used in a point-to-multipoint configuration, wherein an access point communicates via an omnidirectional antenna with one or more nomadic or mobile clients that are located in a coverage area around the access point. Typical indoor range is 30 m (100 ft.) at 11 Mbit/s and 90 m (300 ft.) at 1 Mbit/s. 802.11b cards can operate at 11 bit/s, but will scale back to 5.5, then 2, then 1 Mbit/s (also known as Adaptive Rate Selection), if signal quality becomes an issue [7].

Without loss of generality, the RF front-end design in this paper will focus on the frequency band of 2.4 GHz–2.5 GHz which are used by IEEE standards such as 802.11b and 802.11n [8].

M. Ma (Ed.): ICTCE 2020, LNEE 797, pp. 47–55, 2022.
https://doi.org/10.1007/978-981-16-5692-7_6

2 Design Theories and Target Specifications

The performance of RF front-end for WLAN receivers are tightly regulated by IEEE standards. According to the ongoing IEEE standards such as 802.11b and 802.11n, the minimum specs that the RF receivers should satisfy are shown as Table 1.

Table 1. Minimum specs for WLAN receivers, as regulated by IEEE 802.11b.

Item	Minimum Spec
Noise Figure	10 dB
Gain	40 dB
IIP3	−30 dBm
Channel Bandwidth	20 MHz

The basic design theories of designing a receiver are described as following. Firstly, after the signal is captured by the antenna, a filter is needed to select the band and reject the image band. Because the filter is at the first stage of the system, its noise figure contributed significantly to the overall noise figure. Thus, a filter with smallest possible noise figure is critical to the system's performance. After the filter, the signal needs to be amplified by a low noise amplifier. The amplified signal will be down-converted with a mixer to the IF band. The image frequency signal may introduce some interference to the signal; the image is rejected by the band select filter to reduce the image noise. Another amplifier will be placed after the mixer to further amplify the signal. Finally, a band-pass filter is used to select the target channel.

The block-level diagram of the proposed RF front-end design of the WLAN receiver is shown in Fig. 1.

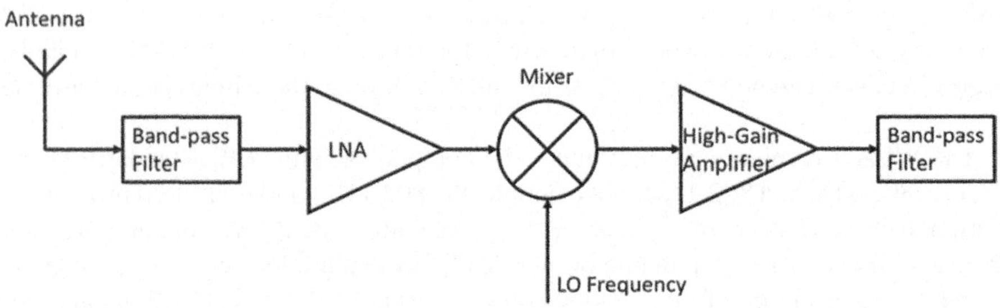

Fig. 1. Block diagram of proposed RF front-end for the WLAN receiver.

3 Design Specifics for Building Blocks

Band-Select Filter. The first filter in Fig. 1 is responsible for choosing the signal band and reject the image frequency signal. So it operates at radio frequency (2.4 GHz–2.5 GHz). Because it is the first stage of the circuit, its noise figure would have a significant influence on the overall noise figure of the receiver. Thus, it is designed for minimum noise figure while having good image frequency rejection.

The design for the band-select filter always face the tradeoff between the noise figure and the out-of-band rejection. For better noise figure of the overall system, a filter with the lowest possible noise figure is preferable, which means lowest in-band attenuation. On the other hand, for reducing the out-of-band interference, we need it to have best possible out-of-band rejection.

There are different structures to design the band pass filter, including coupled line filters, stepped impedance filters, and microstrip stub filters. Also, there are different types of filters such as Chebyshev filter and Maximum Flat filter. The actual design of the filter is done with Advanced Design Systems (ADS) software trying all these options. After these simulations, it is decided that a microstrip coupled line Chebyshev filter is the best choice considering the performance, the cost, and the area. The circuit diagram of this filter can be seen below in Fig. 2.

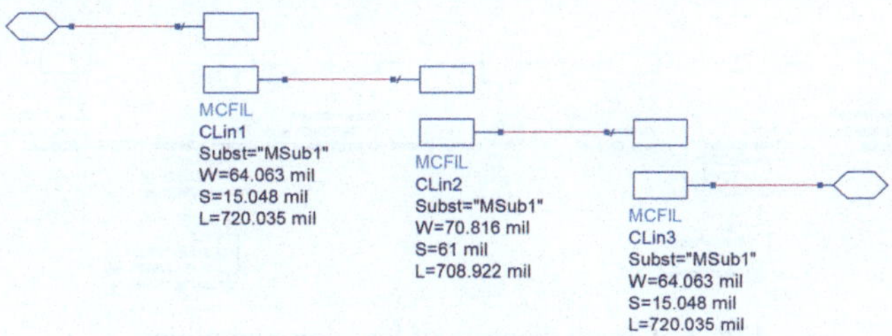

Fig. 2. Circuit diagram of first band-pass filter.

The specifications for the band-select filter are listed in Table 2.

Low Noise Amplifier. The low noise amplifier operates at radio-frequencies with a center frequency of 2.45 GHz. The noise figure for the amplifier is critical to the system's overall noise figure, and thus, this amplifier is designed to have minimal noise. A NESG2031M05 transistor is used as the active device for designing the LNA.

Table 2. Specifications for the band-select filter.

Item	Performance
Lower stop-band frequency	2 GHz
Lower pass-band frequency	2.4 GHz
Higher pass-band frequency	2.5 GHz
Higher stop-band frequency	3 GHz
Type	Chebyshev
Ripples	0.5 dB
Image frequency rejection	30 dB
Pass-band attenuation	1.1 dB
Noise Figure	1.1 dB

A low-noise amplifier with a gain of 15.753 dB, a noise figure of .745 dB and an IP3 intercept point of 10.6356 dB is designed with microstrip line matching network. The circuit diagram of our low-noise amplifier can be seen in Fig. 3.

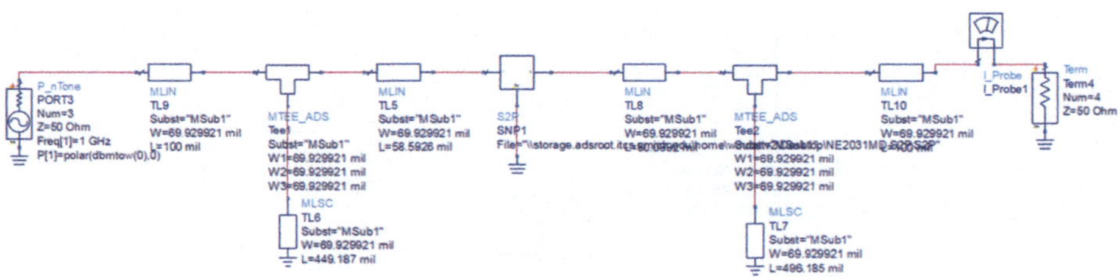

Fig. 3. Circuit diagram of low-noise amplifier.

Table 3 shows a summary of the specifications of the low-noise amplifier.

Table 3. Specifications for the low-noise amplifier.

Item	Performance
Operating frequency	2.45 GHz
Gain	15.753 dB
Noise Figure	0.754 dB
IIP3	10.6356 dB

Mixer. In order to derive the power gain of the mixer, we swept the Local Frequency Input power to find the optimum LO magnitude. Meanwhile, we designed matching circuit between the Wilkinson combiner and the diode, as well as between the diode and the later stages. The matching circuit is designed for conjugate matching and aimed to minimize the power reflection of the RF signal and IF signal. The circuit diagram of our mixer can be seen in Fig. 4.

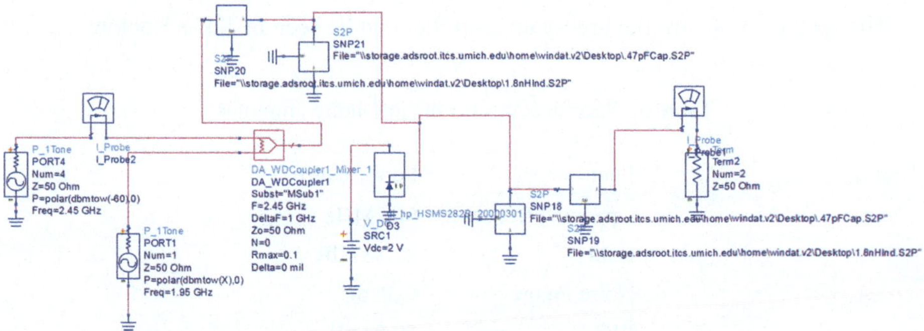

Fig. 4. Circuit diagram mixer.

The specifications for mixer can be seen in Table 4.

Table 4. Specifications for the low-noise amplifier.

Item	Performance
Gain	9.542 dB
Noise Figure	10.0 dB
IIP3	−79.572 dB
High Oscillation Frequency	2.45 GHz
Low Oscillation Frequency	1.95 GHz
Output Frequency	500 MHz

High-Gain Amplifier. The second amplifier is made to operate at IF frequencies. It is designed to have a center frequency of 500 MHz and prioritizes high gain. We designed this amplifier to have maximum gain. In order to get maximum gain, we used the NESG2101M05 transistor.

Lumped elements are used for designing the input and output matching network for the high-gain amplifier as it operates at such a low frequency. The matching network is carefully designed to maximize the gain while maintaining stability. According to simulation results, a small resistor is on the output matching network to ensure stability.

The circuit diagram of the designed high-gain amplifier can be seen in Fig. 5.

Fig. 5. Circuit block diagram of high-gain amplifier.

The specifications for the high-gain amplifier can be seen in Table 5 below.

Table 5. Specifications for the low-noise amplifier.

Item	Performance
Operating frequency	500 MHz
Gain	27.337 dB
Noise Figure	0.35 dB
IIP3	10.64 dB

Channel-Select Filter. The second filter is needed for distinguishing the different channels. Thus the bandwidth is the channel bandwidth which is 20 MHz. Because the channels are closely next to each other, we need a filter with fast attenuation at its stop-band frequencies. Meanwhile, because we set the IF to be 500 MHz, the fractional bandwidth is $20/500 = 4\%$, which is relatively small. In addition, if we design the filter with microstrip lines, the filter will not work well in rejection at some of its harmonic frequencies because of the periodic character of microstrip lines. Meanwhile, designs with lumped elements are often not feasible due to the availability of components with the needed L values and/or C values.

In order to solve the design challenges of the second band-pass filter, we used a two-stage filter, which is the series connection of a microstrip line filter and a lumped component filter. The first one, which is consisted of microstrip lines, can provide a fast attenuation at the stop band and lower down the requirements for the lumped-element filter, and the second one, which is consisted of lumped L and C, can work to reject the signals at harmonic frequencies. The circuit diagram of the second band-pass filter can be seen in Fig. 6.

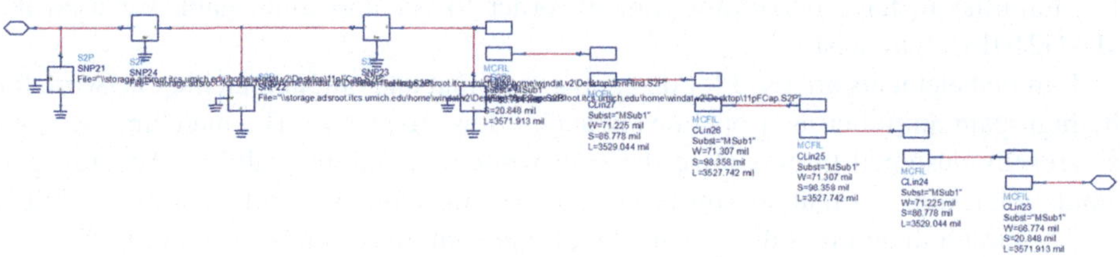

Fig. 6. Circuit block diagram of second band-pass filter.

A summary of the specifications for the second band-pass filter can be seen in Table 6.

Table 6. Specifications for the low-noise amplifier.

Item	Performance
Lower stop-band frequency	480 MHz
Lower pass-band frequency	490 MHz
Higher pass-band frequency	510 MHz
Higher stop-band frequency	520 MHz
Type	Two Chebyshev filters in series
Pass-band Ripples	0.5 dB
Pass-band Attenuation	6.887 dB
Noise Figure	6.887 dB

4 System Level Simulation Results

System level simulation results show that the proposed design method achieves very good performance including the noise figure, the gain, and the IIP3. The summary of our design specifications can be seen below in Fig. 7, Fig. 8, Fig. 9 and as summarized in Table 7.

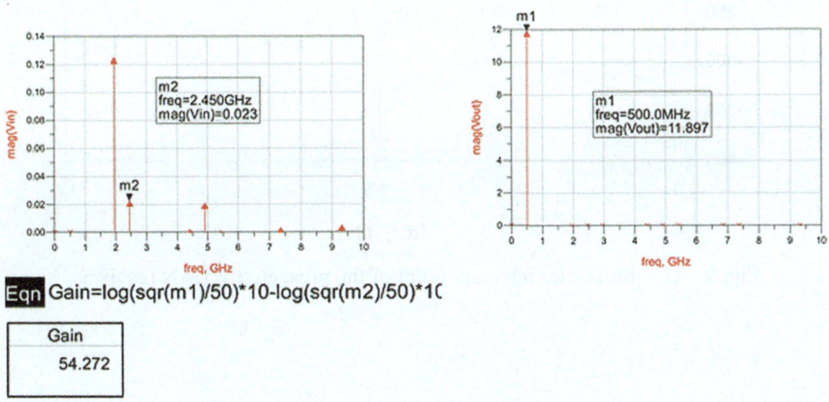

Fig. 7. The gain of the presented WLAN receiver.

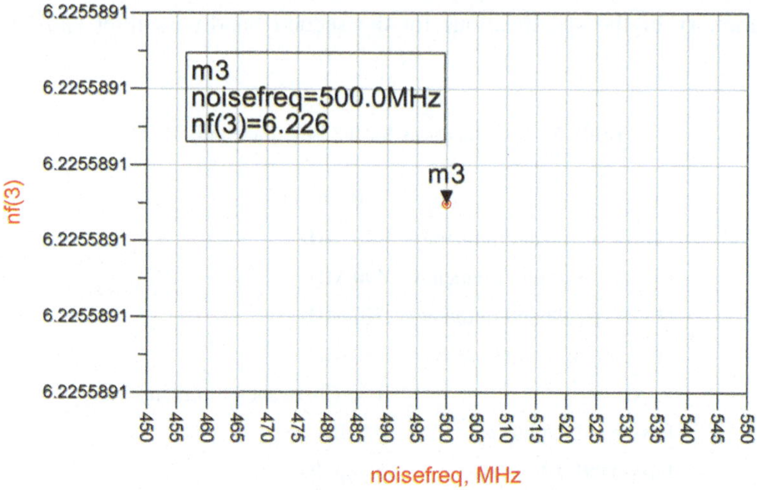

Fig. 8. The noise figure of the presented WLAN receiver.

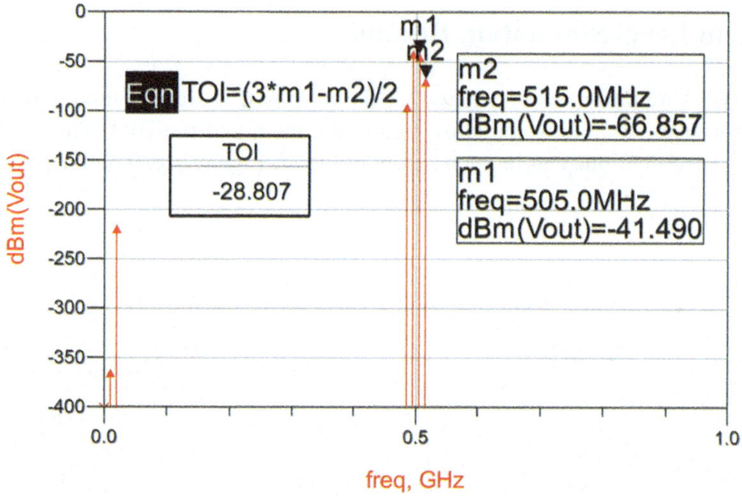

Fig. 9. The third order intercept point of the presented WLAN receiver.

Table 7. The performance of the WLAN receiver.

Item	Performance
Receiver Sensitivity	−123.0 dB
Gain	54.272 dB
Noise Figure	7.94 dB
Dynamic Range	42.8 dB
IIP3	−28.807 dB

5 Conclusion

In this paper, we designed the front-end of a WLAN receiver. Receiver architecture, circuit design and simulation results are demonstrated. Our design can down-convert a typical WLAN signal from 2.4–2.5 GHz to an IF frequency of 500 MHz, and the channel bandwidth is designed as 20 MHz. The proposed design has a gain of 54.272 dB a noise figure of 7.94 dB. The IIP3 of the receiver is −28.807 dB. Receiver sensitivity is −123 dB, while the dynamic range is 42.8 dB. The specifications of the receiver satisfy the standard of commercial WLAN applications.

References

1. Khalajmehrabadi, A., Gatsis, N., Akopian, D.: Modern WLAN fingerprinting indoor positioning methods and deployment challenges. IEEE Commun. Surv. Tutor. **19**(3), 1974–2002 (2017)
2. Xiaomu, H., Yan, S., Vandenbosch, G.A.: Wearable button antenna for dual-band WLAN applications with combined on and off-body radiation patterns. IEEE Trans. Antennas Propag. **65**(3), 1384–1387 (2017)
3. Deng, J., Li, J., Zhao, L., Guo, L.: A dual-band inverted-F MIMO antenna with enhanced isolation for WLAN applications. IEEE Antennas Wirel. Propag. Lett. **16**, 2270–2273 (2017)
4. Abdulraheem, Y.I., et al.: Design of frequency reconfigurable multiband compact antenna using two PIN diodes for WLAN/WiMAX applications. IET Microwaves Antennas Propag. **11**(8), 1098–1105 (2017)
5. Ali, T., Biradar, R.C.: A triple-band highly miniaturized antenna for WiMAX/WLAN applications. Microw. Opt. Technol. Lett. **60**(2), 466–471 (2018)
6. Chieh-Pin, C.: A high gain, low noise WLAN receiver for dual if double down-conversion application in 90-nm RF CMOS. Microw. Opt. Technol. Lett. **49**(10), 2422-2425 (2007)
7. Pozar, D.: Microwave Engineering, 3rd edn., John Wiley and Sons Inc., Hoboken, NJ (2005). IEEE Criteria for Class IE Electric Systems (Standards style), IEEE Standard 308, 1969
8. IEEE: Standard for Information Technology, IEEE Standard 802.11 (1999)

Parallel Optical Wireless Communication System with Hierarchical Nonorthogonal Code Shift Keying

Nobuyoshi Komuro[1(✉)] and Hiromasa Habuchi[2]

[1] Chiba University, 1-33 Yayoi-cho, Inage-ku, Chiba-shi, Chiba 263-8522, Japan
kmr@faculty.chiba-u.jp

[2] Ibaraki University, 4-21-1 Nakanarusawa, Hitachi 316-8511, Ibaraki, Japan

Abstract. This paper proposes a scheme for enhancing data transmission efficiency of Intensity Modulation Direct Detection (IM/DD) optical wireless communications (OWCs) by combining the parallel OWC with hierarchical nonorthogonal code based multilevel modulation. In the proposed system, nonorthogonal codes are constructed by concatenating pseudo orthogonal M-sequences. The numerical results show that the data transmission efficiency of the proposed system can achieve 1.0, which shows the effectiveness of the combination of the parallel OWC and the hierarchical nonorthogonal code based multilevel modulation system.

Keywords: Code shift keying · Intensity Modulation Direct Detection (IM/DD) · Parallel optical wireless communication

1 Introduction

In the last few decades, wireless communication technology has improved and spread dramatically [1–26]. In particular, Optical Wireless Communications (OWCs) have been focused on for the past decade. The Intensity Modulation Direct Detection (IM/DD) OWC system have been investigated as practical OWCs.

The Sequence Inversion Keying (SIK) system is one of the information modulation systems in the IM/DD OWCs. It is relatively simple for IM/DD OWCs to implement SIK system since no threshold setting is required. However, the data transmission efficiency of the SIK system is not high since the SIK system transmits a bit per symbol. It is desirable to increase the data transmission efficiency of the IM/DD OWCs.

In [26], the Code Shift Keying (CSK) system has been investigated for increasing the data transmission efficiency. In the CSK system, a transmitter selects one of M codes based on the source data. The CSK system improves the bits per code by increasing the number of codes. It has been reported that the data transmission efficiency could be enhanced by using parallel combinatory communication [28] and also hierarchical modulation schemes [30]. In addition, the authors have reported that the data transmission efficiency of the CSK system could be improved by concatenating CSK codes [27].

M. Ma (Ed.): ICTCE 2020, LNEE 797, pp. 56–65, 2022.
https://doi.org/10.1007/978-981-16-5692-7_7

This paper proposes a method for enhancing the data transmission efficiency of the IM/DD OWCs by combining the parallel OWC system, hierarchical modulation scheme, and the nonorthogonal CSK system. In the proposes system, CSK codes are constructed by concatenation Pseudo Orthogonal M-Sequences (POMs). This paper derives the data transmission efficiency of the proposes system in the single user case. The numerical results show the effectiveness of the proposed system.

2 System Construction

2.1 Construction of a Nonorthogonal Code

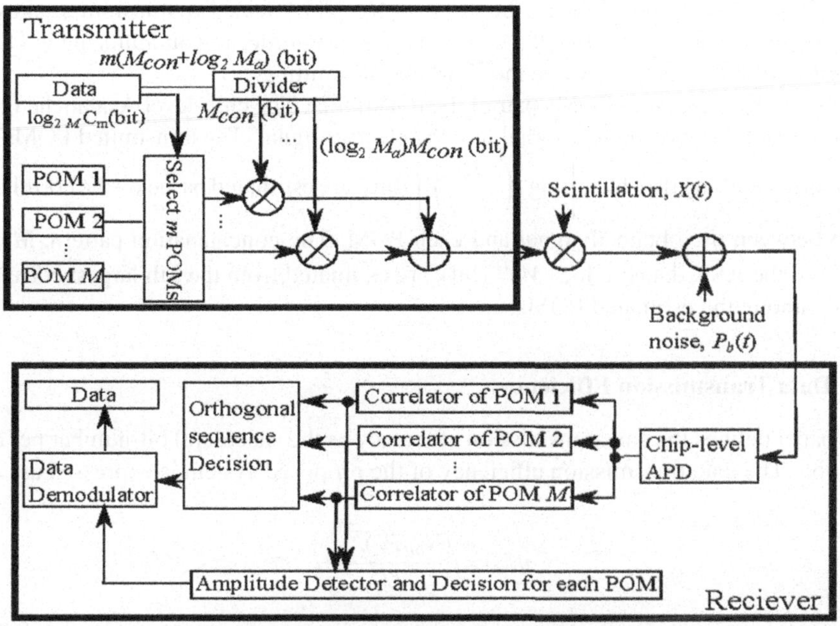

Fig. 1. Transmitter and receiver model.

A proposed nonorthogonal code is composed by concatenating M_{con} Pseudo Orthogonal M-sequences (POMs). POM pair sets [29] are used as primitive codes. In the proposed scheme, m of M_{oc} POM codes are selected from $\left\lfloor \log_2 \binom{M_{oc}}{m} \right\rfloor$ (bit) data and each of them are concatenated by $M_{con}{}^{(i)}$ data, where M is the POMs number and $M_{con}{}^{(i)}$ is the concatenation pattern of i-th selected POM. In addition, each primitive code is modulated by $M_a{}^{(i)}$ Amplitude Modulation Keying (ASK) system, where $M_a{}^{(i)}$ is the number of amplitude levels. The ASK modulation stage can be regarded as the parallel combinatory communication of concatenating patterns. A frame has $\left\lfloor \log_2 \binom{M_{oc}}{m} \right\rfloor$

$+\, m(M_{con} + \log_2 M_a^{(i)})(= N_{bit})$ (bit) data. A frame length, L_f, is $L_{oc}M_{con}$, where L_{oc} is the length of a primitive code.

2.2 Transmitter and Receiver Operations

Figure 1 shows the proposed-system model. The transmitter prepares POMs with $\{+1, 0\}$-value. The transmitter modulates the source data according to $\left\lfloor \log_2 \binom{M_{oc}}{m} \right\rfloor$ $+\, m(M_{con} + \log_2 M_a)(= N_{bit})$ (bit) data. m of M POM codes are selected according to $\left\lfloor \log_2 \binom{M_{oc}}{m} \right\rfloor$ (bit) data. The transmitter concatenates each selected POM according to the polarity of ${M_{con}}^{(i)}$ (bit) data. In addition, the transmitter performs ASK modulation for each selected POM code according to $\log_2 M_a^{(i)}$ (bit) data. The transmitter takes the sum of concatenated codes by chip and transmits it. In order to communicate in OWC, the transmitter converts -1 value into 0 when transmitting.

The receiver prepares POMs pair of the transmitter. The chip-level Avalanche Photo Diode (APD) converts optical signal into the electric signal. The transmitted POMs and their corresponding data bits, $\left\lfloor \log_2 \binom{M_{oc}}{m} \right\rfloor$ (bit), are estimated based on the correlation value between the obtained signal and each POM. The concatenation pattern, ${M_{con}}^{(i)}$ (bit), and the ASK data bit, $\log_2 M_a^{(i)}$ (bit), are estimated from the i-th largest correlator output among the estimated POMs.

2.3 Data Transmission Efficiency

This paper defines the data transmission efficiency as the successful bit-number per chip duration. The data transmission efficiency of the proposed system is expressed as

$$\eta_{sys} = \frac{(N_{bit}P_c)^{\frac{L_{bit}}{N_{bit}}}}{L_f},$$

where P_c is the symbol success rate and L_{bit} is the bit number per packet.

3 Transmission Efficiency Analysis

This paper analyzes the data transmission efficiency in the single user case. This paper assumes background, APD, and thermal noises, and the scintillation. The notation on the following analysis is shown in Table 1.

Table 1. Notations

T_c	Chip length
K_{eff}	APD ionization ratio
G	Gain of the APD
F	Excess noise figure
Me	Modulation extinction ratio
η	Efficiency of quantum
k_B	Boltzmann's constant
T_r	Noise temperature at the receiver
R_L	Resistor at the receiver
e	Electricity quantum
σ_{th}^2	Thermal noise variance
σ_s^2	Scintillation-logarithm variance
P_b, P_w	Background-noise and receiver laser powers w/o scintillation
h	Planck's constant
λ_s	$\left(\frac{\eta P_w}{hf}\right)$
λ_b	$\left(= \frac{\eta P_w}{hf}\right)$
I_b, I_s	APD bulk and surface leakage currents

The symbol success ratio, P_c, is

$$P_c = m\int_0^\infty P(X)\left\{1 - \frac{1}{M_a}\left[\sum_{i=1}^{M_a-1}\frac{1}{2}erfc\left(-\frac{Th_{i-1}-\mu_i}{\sqrt{2\sigma_s^2}}\right)\right.\right.$$
$$\left.\left.+\sum_{i=1}^{M_a-1}\frac{1}{2}erfc\left(\frac{Th_i-\mu_i}{\sqrt{2\sigma_s^2}}\right)+\sum_{i=1}^{M_a-1}\frac{1}{2}erfc\left(-\frac{Th_{M_a}-\mu_{M_a}}{\sqrt{2\sigma_s^2}}\right)\right]\right\}$$
$$\times\,(1-P_{oc}(X))dX, \tag{1}$$

where μ_0 and σ_0^2 are the average and variance for POM polarity, and *P(X)* is the probability density function (PDF) of optical intensity in consideration of the scintillation *X* [23], which are

$$\mu_0(\mathrm{X}) = \mathrm{G}T_c\left(\frac{L_f+1}{2}\lambda_s X - \frac{L_f-1}{2}\frac{\lambda_s X}{Me}\right), \tag{2}$$

$$\sigma_0^2(X) = G^2FT_c\left[m\frac{L_f+1}{2}\lambda_s X + m\frac{L_f^2-1}{2}\frac{\lambda_s X}{Me} + L_f\lambda_b + \frac{2I_b}{e}\right] + \frac{2I_sT_c}{e} + 2\sigma_{th}^2, \tag{3}$$

and

$$P(X) = \frac{1}{\sqrt{2\pi\sigma_s^2}} \exp\left[-\frac{\left(\ln X + \frac{\sigma_s^2}{2}\right)^2}{2\sigma_s^2} \right],$$

respectively.

In Eq. (1), P_{oc} is the symbol error rate of a POM code. If a POM code c_1 is sent, P_{oc}, is

$$P_{oc}(X) = 1 - \int_{-\infty}^{\infty} f(c_1) \left[\int_{-\infty}^{c_1} \underset{i>m}{f(c_i, X)\, dc_i} \right]^{M_{oc}-m} \left[\int_{c_1}^{\infty} f(c_m, X) dc_m \right]^{m-1} dc_1, \quad (4)$$

where $f(c_j)$ is the PDF of correlation output for j-th POM c_j, which is

$$\mathrm{f}(c_j, \mathrm{X}) = \mathrm{g}\left(\left|q_j^{(1)}\right|, X\right) * \cdots * \mathrm{g}\left(\left|q_j^{(M_{con})}\right|, X\right), \quad (5)$$

$$\mathrm{g}(q_j, X) = \frac{1}{\sqrt{2\pi\sigma_j^2(X)}} exp\left[-\frac{\left(q_j - \mu_j(X)\right)^2}{\sqrt{2\sigma_j^2(X) M_{con}}} \right]. \quad (6)$$

"*" expresses the convolutional integral, μ_j and σ_j^2 are the average and variance for j-th correlation output which are

$$\mu_1(\mathrm{X}) = \cdots = \mu_m(X) = GT_c\left(\frac{L_f + 1}{2}\lambda_s X + \frac{L_f - 1}{2}\frac{\lambda_s X}{Me} + \lambda_b \right), \quad (7)$$

$$\mu_{m+1}(\mathrm{X}) = \cdots = \mu_{M_{oc}}(X) = GT_c\left(\frac{L_f \lambda_s X}{Me} + \lambda_b \right), \quad (8)$$

and

$$\sigma_s^2(\mathrm{X}) = \cdots = \sigma_s^2(X) = G^2 F T_c \left[m\frac{L_f + 1}{2}\lambda_x X + m\frac{L_f^2 - 1}{2}\frac{\lambda_s X}{Me} + L_f \lambda_b + \frac{2I_b}{e} \right] + \frac{2I_s T_c}{e} + 2\sigma_{th}^2, \quad (9)$$

respectively. In Eqs. (7)–(9), F and σ_{th}^2 are

$$\mathrm{F} = k_{eff} G + \left(1 - k_{eff}\right)\frac{2G - 1}{G},$$

and

$$\sigma_{th}^2 = \frac{2k_B T_r T_c}{e^2 R_L},$$

respectively. The data transmission efficiency is

$$\eta_{sys} = \frac{(N_{bit} P_c)^{\frac{L_{bit}}{N_{bit}}}}{L_f}, \quad (10)$$

where L_{bit} is the bit number per packet.

4 Numerical Results

Table 2 presents the evaluation parameters. Figure 2 presents the data transmission efficiency vs. the transmit laser power per bit, P_w, of the conventional CSK and the proposed system without ASK modulation when L_{bit} is 4000 (bit) and the number of combinatory codes, m, is 2. In the proposed system, the combinations of (M_{oc}, M_{con}) are $(7, 3)$, $(7, 4)$, $(15, 2)$, and $(31, 1)$. The frame length, the number of POMs, and the number of bits per frame of the conventional CSK are 31 (chip), 31, and $\left\lfloor \log_2 \binom{M_{oc}}{m} \right\rfloor$ (bit) respectively. In the proposed system, the frame lengths are 21, 28, 30, and 31 (chip) for $(M_{oc}, M_{con}) = (7, 3)$, $(7, 4)$, $(15, 2)$, and $(31, 1)$, respectively. From Fig. 2, the proposed system with $(M_{oc} = 7, M_{con} = 3)$ achieves the highest data transmission efficiency.

Table 2. Evaluation parameter

Parameter	Value
Laser wavelength	830 (nm)
T_c	4.0×10^{-4}(μsec)
G	100
P_b	−45.0 (dBm)
I_b	0.1 (nA)
I_s	10 (nA)
Me	100
T_r	300 (K)
R_L	1030 (Ω)
σ_s^2	0.1
E	1.6×10^{-19}(C)
η	0.6
k_{eff}	0.02

Figure 3 presents the data transmission efficiency vs. the number of combinatory codes, m, of the conventional CSK and the proposed systems without hierarchical modulation when L_{bit} is 4000 (bit). In Fig. 3, we assume the ideal case. In other words, we do not assume background noise, thermal noise, and scintillation. From Fig. 3, the data transmission efficiencies of the proposed system can exceed 1.0 under the ideal case while that of the conventional CSK does not achieve 1.0. In addition, from Fig. 3, the proposed system with ($M_{oc} = 31$, $M_{con} = 1$) achieves the highest data transmission efficiency under the ideal case.

Figure 4 presents the data transmission efficiency vs. the number of combinatory codes, m, of the conventional CSK and the proposed systems with hierarchical modulation when L_{bit} is 4000 (bit). In the proposed system, the combinations of (M_{oc}, M_{con}, M_a) are (7, 3, 4), (7, 4, 4), (15, 2, 2), and (31, 1, 1). In Fig. 4, we assume the ideal case. In other words, we do not assume background noise, thermal noise, and scintillation. From Fig. 4, the data transmission efficiencies of the proposed system can exceed 1.0 under the ideal case while that of the conventional CSK does not achieve 1.0. Also, From Fig. 3, the proposed system with ($M_{oc} = 7$, $M_{con} = 3$, $M_a = 4$) achieves the highest data transmission efficiency under the ideal case, which shows the effectiveness of the combination of the nonorthogonal CSK, parallel OWC, and hierarchical modulation scheme.

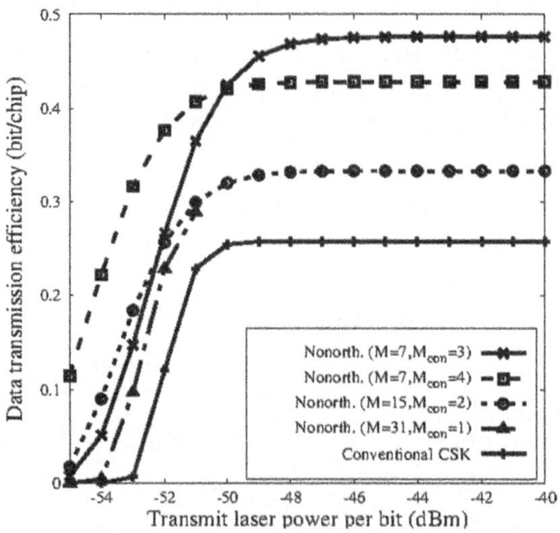

Fig. 2. Data transmission efficiency of the CSK systems ($Lbit$ = 4000 (bit) and $P_w = -48.0$ (dBm)).

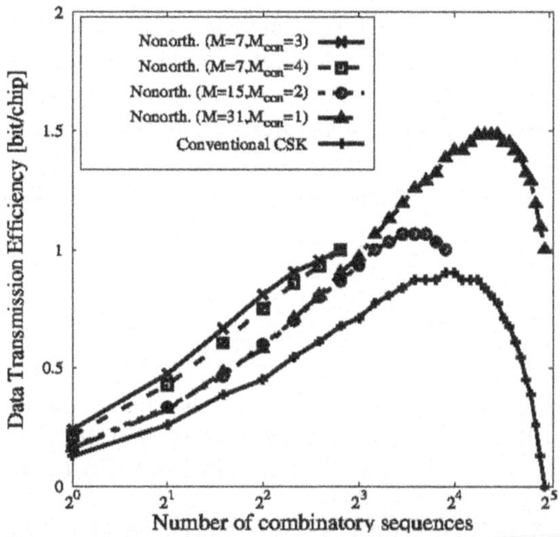

Fig. 3. Data transmission efficiency of the CSK systems without hierarchical modulation under ideal case (*Lbit* = 4000 (bit)).

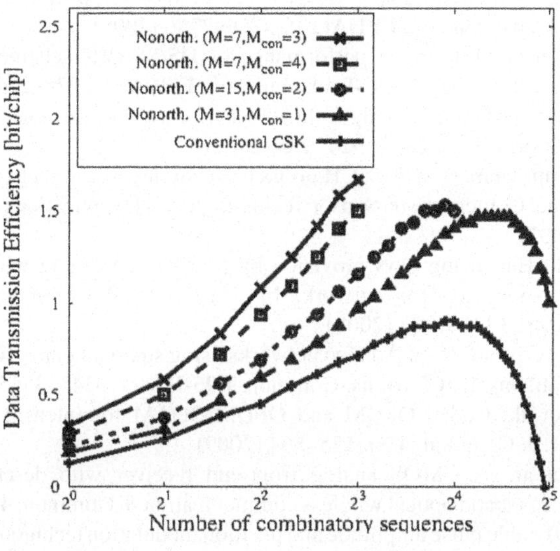

Fig. 4. Data transmission efficiency of the CSK systems with hierarchical modulation under ideal case (*Lbit* = 4000 (bit)).

5 Conclusion

This paper proposed a method for enhancing the data transmission efficiency of the IM/DD OWCs by combining the parallel OWC system, hierarchical modulation scheme, and the nonorthogonal CSK system. In the proposes system, CSK codes are constructed by concatenating pseudo orthogonal M-sequences. This paper derived the data transmission efficiency of the proposes system in the single user case. The numerical results showed that the data transmission efficiency of the proposed system could achieve 1.0, which showed the effectiveness of the hierarchical nonorthogonal CSK system.

Acknowledgement. This work was supported by the Telecommunications-ADVANCEMENT Foundation.

References

1. Kwon, H.M.: Optical orthogonal code-division multiple-access system; part I: APD noise and thermal noise. IEEE Trans. Commun. **42**(7), 2470–2479 (1994)
2. Tseng, S.M.: A High-throughput multicarrier DS CDMA/ALOHA network. IEICE Trans. Commun. **E86-B**(4), 1265–1273 (2003)
3. Komuro, N., et al.: CSK/SSMA ALOHA system with nonorthogonal sequences. IEICE Trans. Fundam. **E87-A**(10), 2564–2570 (2004)
4. Komuro, N., et al.: A reasonable throughput analysis of the CSK/SSMA unslotted ALOHA system with nonorthogonal sequences. **E88-A**(6), 1462–1468 (2005)
5. Komuro, N., et al.: Nonorthogonal CSK/CDMA with received-power adptive access control scheme. IEICE Trans. Fundam. **E91-A**(10), 2779–2786 (2008)
6. Kobayashi, K., et al.: Improving performance of DS/SS-IVC scheme based on location oriented PN code allocation. IEICE Trans. Fundam. **E99-A** (1), 225–234 (2016)
7. Lindsey, S., et al.: Data gathering algorithms in sensor networks using energy metrics. IEEE Trans. Parallel Distrib. Syst. **13**(9), 924–935 (2002)
8. Fan, X., et al.: Improvement on LEACH protocol of wireless sensor network. In: Proceedings of the International Conference on Sensor Technologies and Applications (SENSORCOMM), pp. 260–264 (2007)
9. Luo, C.Y., et al.: Enhancing QoS provision by priority scheduling with interference drop scheme in multi-hop Ad Hoc network. In: IEEE Global Communication Conference (GLOBECOM), pp. 1321–1325 (2008)
10. Ma, J., et al.: MAC protocol for Ad Hoc networks using smart antennas for mitigating hidden and deafness problems. IEICE Trans. Commun. **E95-B** (11), 3545–3555 (2012)
11. Tabach, M.E., et al.: Coded OFDM and OFDM/OQAM for intensity modulated optical wireless systems. J. Commun. **4**(8), 555–564 (2009)
12. Joyner, V.M., et al.: A CMOS analog front-end receiver with desensitization to input capacitance for broadband optical wireless communication. J. Commun. **4**(8), 572–581 (2009)
13. Zeng, Y., et al.: Tunable pulse amplitude and position modulation technique for reliable optical wireless communication channels. J. Commun. **2**(2), 22–28 (2007)
14. Fuada, S., et al.: Design of reconfigurable system-on-chip architecture for optical wireless communication. J. Commun. **14**(10), 965–970 (2019)
15. Qiu, Y., et al.: Visible light communication based on CDMA technology. IEEE Wirel. Commun. **25**(2), 178–185 (2018)
16. Liu, M.Y., et al.: Throughput performance analysis of asynchronous optical CDMA networks with channel load sensing protocol. IEEE Photonics J. **9**(3), 1–13 (2017)
17. Chen, S.H., et al.: Color-shift keying code-division multiple-access transmission for RGB-LED visible light communications using mobile phone camera. IEEE Photonics J. **7**(6), 1–6 (2014)

18. Kiasaleh, K.: Performance of APD-based, PPM free-space optical communication systems in atmospheric turbulence. IEEE Trans. Commun. **53**(9), 1455–1461 (2005)
19. Zhu, X., et al.: Free-space optical communication through atmospheric turbulence channels. IEEE Trans. Commun. **50**(8), 1293–1300 (2002)
20. Hadi, M., et al.: Analysis and design of adaptive OCDMA passive optical networks. J. Lightwave Technol. **35**(14), 2853–2863 (2017)
21. Li, J., et al.: Optical wireless communications: system model, capacity and coding. In: Proceedings of the IEEE VTC, vol.1, pp.168–172 (2003)
22. Yamashita, T., et al.: An Optical Code Division Multiplexing System using Hadamard Codes and SIK. IEICE Technical report, OCS98-1 (1998)
23. Kozawa, Y., et al.: Theoretical analysis of M-CSK/CDMA system in optical wireless channel. In: Proceedings of the International Symposium on Information Theory and Its Applications (ISITA), pp.738–742 (2007)
24. Ochiai, N., et al.: Performance analysis of synchronous optical CDMA system with EWO signaling. IEICE Trans. Fundam. **J86-A**(9), 957–968 (2003)
25. Hsieh, C.P., et al.: A bipolar-bipolar code for asynchronous wavelength-time optical CDMA. IEEE Trans. Commun. **54**(7), 2572–2578 (2006)
26. Takamaru, Y., et al.: Theoretical analysis of new PN code on optical wireless code-shift-keying. IEICE Trans. Fundam. **E97-A**(12), 2572–2578 (2014)
27. Komuro, N., et al.: Intensity modulation direct detection optical wireless communication with nonorthogonal code shift keying. In: Proceedings of the IEEE Global Conference on Consumer Electronics (GCCE), pp.723–726 (2019)
28. Zhu, J., et al.: Proposal of parallel combinatory spread spectrum communication system. IEICE Trans. Commun. (Jpn. Ed.) **J74-B**(5), 207–214 (1991)
29. Habuchi, H.: Pseudo-noise sequences based on M-sequence and its application for communications. IEICE Fund. Rev. (Jpn. Ed.) **3**(1), 32–42 (2009)
30. Tokunaga, T., et al.: New two-layered pseudo-noise code for optical-wireless code-shift keying/SCDMA. In: Proceedings of the International Workshop on Signal Design and Its Applications in Communications (IWSDA), pp.149–153 (2017)

JomIoT Medic: Saving Lives

Z. A. Atallah[1], P. S. JosephNg[1(✉)], and Y. F. Loh[2]

[1] Institute of Computer Science and Digital Innovation, UCSI University, UCSI Heights, 56000 Cheras, Kuala Lumpur, Malaysia
josephng@ucsiuniversity.edu.my

[2] Faculty of Business and Management, UCSI University, UCSI Heights, 56000 Cheras, Kuala Lumpur, Malaysia
lohyf@ucsiuniversity.edu.my

Abstract. Nonadherence medication refers to the misuse or forgetting to take medication on time, it Is a fatal issue that causes a huge number of health issues and complex failure in the body and unfortunately. This paper proposes an advanced innovative medication reminder device consist of a mobile application and it is connected wirelessly using IoT to the hardware part as a special medication container in a shape of a smartphone cover or chain. To evaluate the device a survey was distributed to different population samples of the different age groups from both local and private sectors followed by an intense interview from the same sampling population. This medication device is the perfect and most convenient method to stick to medication time and the number of pills to help in fast recovery at all ages without complications and most importantly to save money. Managerial and practical implications are discussed.

Keywords: IoT · Medication adherence · Wireless connection · Smartphone cover

1 Medication Adherence and Medication Reminder Device

Medications, in general, has to be taken on time, however, it is estimated that almost 50% of the population forget to take their medication at least once a month [1]. Forgetting the medicine is the most frequently reported reason for not taking it on time, it subsequently affects the recovery or in severe cases, it can be life-threatening especially for those who have chronic diseases with high blood pressure, high cholesterol or diabetes and forgetting the required medication can then increase the risk of heart attack, kidney failure or other diseases that lead to disability, morbidity, or mortality [2–4]. On the other hand, forgetting to consume the antibiotics on time can decrease the cure rate and increase the possibilities of developing more bad bacteria [5, 6], where the patients need to visit their doctors more often.

Misuse or forgetting medication has a negative effect on the cost as well as the pressure on the health section. Globally, a huge amount of money around 290$ billion annually is wasted on fixing medication adherence errors [7]. This paper proposes an innovative medication reminder device that significantly improves medication adherence

M. Ma (Ed.): ICTCE 2020, LNEE 797, pp. 66–77, 2022.
https://doi.org/10.1007/978-981-16-5692-7_8

in all ages which twill hen has a positive effect on the health section. To promote and build an efficient mechanism of medication reminder device the connection between the medication reminder application and the special container is made using the internet of things (IoT) [8], to remotely control the case from the app. The remainder will have information of all medicines in the container including the time of consumption and the number of pills to be taken at each time, besides the application can control the time of medication whether it is before or after meals or the amount of water a patient should drink together with the medicine.

The significance and benefits of this study are to develop and positively change the overall health system in the world toward a healthier sickness free environment by increasing the medication adherence which can improve the health and safety as it is designed to assist people of all ages to take their pills on time while avoiding errors and without the need of a third party or a complicated process. These are hypotheses for each point of it examined in Table 1, acknowledged from the medication reminder device, and summarized in Fig. 1.

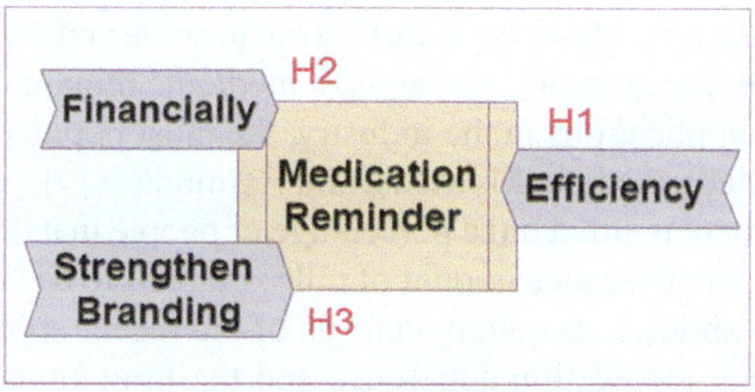

Fig. 1. Hypothesis model design

Table 1. Research hypothesis

H1: MedicMinder contribute to saving cost for the health section
H2: MedicMinder ensure medication adherence
H3: MedicMinder establish trust within the users and application

Based on Table 1 there are three hypotheses related to cost, efficiency, and brand trust, that are required and encouraged to strike from this study.

H1: Implementing this innovative method of medication reminder will contribute to saving the efforts as well as the cost for the health section which results in significantly improving the overall health of individuals.

H2: The special container must be innovative and unique by taking the shape of a smartphone cover or chain instead of the traditional individual dispenser to ensure that users never forget the container which by default increases the medication adherence significantly.

H3: To establish significant trust within the users and application, when patient press 'Take Medicine' to take the medication, the phone case will not slide the medicine before requesting the patient code which will be given to the patient upon registering for the application or it could be done as a fingerprint or face recognition depends on the phone model as well as the patient request during the application installation.

1.1 Value Creations

This medication reminder innovative device significantly improves the user's overall health. Allowing users to experience a modern convenient way that positively increases their medicine compliance which will save their lives and assure a better treatment. The novelty of this developed medication reminder device is the use of IoT to connect two parts which are the software (reminder application) and the innovative smart idea of the hardware part being the smartphone cover or chain, to contribute in a positive efficient way in the industry of medication adherence.

2 Legacy and Related Works of Medication Reminders

Many attempts and researches have been carried out to solve and maintain the problem of medication adherence, for example, setting up a medicine remainder application. There are many alarm base applications in the industry, the most popular medicine remainder applications are MyMedSchedule, MyMeds, and RxmindMe [9], yet the problem is still not fully solved and there is still a huge percentage of people that do not take medication on time and never stick to the exact amount of pills or tablets to be taken [10]. The related works Table 2 below shows a thorough analysis of the recent approaches and methods dedicated to increasing medication adherence and the limitations or disadvantages of each.

Fig. 2. An example of the non-smart medication case

3 Methodology

In this study, a smart medicine container is proposed. This container consists of a mobile application as a medicine remainder and monitor, the unique smart architecture of this

Table 2. Related work review

No	Title	Method of reminder	Disadvantages
1	Comprehensive Approach for A Smart Medication Dispenser [11]	The Smart Medication Dispenser (SMD) builds and connected to a software application using Bluetooth, to increase medication adherence	Pill dispenser is too big to have remote features The prototype works by a bit of movement in the pill dispenser which results in damage in the long term
2	Medication remainder and healthcare _ an android application [12]	Focus on connecting patients with doctors using an android mobile application, Use an alarm, and notification reminder	Not for general use, because it's only suitable for patients under doctor's observation Does not provide video chatting or online consultation with doctors
3	A Smart Pill Box to Remind of Consumption using IoT [13]	Consist of a pillbox that holds the medication and a real-time clock, remind users by sending messages or email	The complicated procedure of writing the number of pills taken after each reminder
4	Smart drugs: Improving healthcare using Smart Pill Box for Medicine Reminder and Monitoring System [14]	Uses a pillbox connected wirelessly to a mobile application. Pillbox has a special design of dispensing the pills with 9 units	It is not fully advanced as it requires manual procedures for inserting pills as well as running the system
5	Smart pill dispenser using the Internet of Things [15]	Remind the user via GSM to consume medication. The pills are placed in a smart pillbox and controlled using ultrasound sensors	Do not provide standard details about medication as users have to manually fill up all details regarding the medications and their number
6	Mobile Apps for Increasing Treatment Adherence: Systematic Review [16]	In this study, an analysis study was conducted to analyze the methods and operation of multiple medication methods	Limitations in the database affected the evaluation to be unprecise Did not consider an accurate solution to completely fix the medication adherence issue
7	A medication reminder mobile app: does it work for different age range [17]	Mobile medical app on the Android platform that is appropriate for those who need support with drug regimens based on the age and the need for this application	The evaluation in this application was done by the developers, therefore the results might not be accurate and generalized to all applications and medication reminder methods
8	Medicine Reminder and Monitoring System for Secure Health Using IoT [18]	An analysis is conducted for health medication and monitoring Reminder connected using the internet to view screen	The need to develop a monitoring device of the display screen, instead of taking advantage of the mobile phone screen
9	IoT based Advanced Medicine Dispenser Integrated with an Interactive Web Application [19]	Pill dispenser is developed to increase and improve medication adherence using IoT, cloud computing, and machine learning	Based on the test conducted the application had a few disadvantages: Some users forgot to take the medication dispenser Unable to store liquid medications
10	Demo Abstract: Mobile Sensing to Improve Medication Adherence [20]	Use activity learning application (AL) to study human activities for better reminding	It does not provide a convenient way of carrying the system remotely, therefore this consequence in issues of medication adherence

device is illustrated in Fig. 3. The container is in the shape of a phone cover or chain, which is very convenient for easy accompany wherever we are. The reminder application is connected wirelessly based on the Internet of Things to the container.

This method overcomes the issues and disadvantaged of previous works by using a remote method based on the assumption that nowadays almost everyone in the world has a mobile phone or keychain and they never forget to take either one with them wherever they go [21–25]. The medicines for instant are kept in the phone cover inside the container, which will have a specific shape that is not too big and not too small, yet it is just nice to be part of a phone cover or chain. This container made of a material that

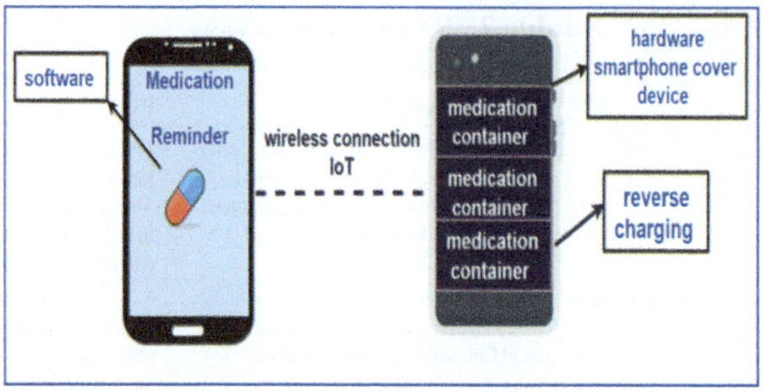

Fig. 3. Diagram of the device

is suitable to store the pills and it is waterproof and not affected by sunlight, radiation, or outside temperature.

This method of medication reminder will be evaluated by applying a method of data collection and analysis (Fig. 4). The data collection method is conducted online through a mixed-mode evaluation (as shown in Fig. 4) based on quantitative and qualitative research findings to improve the conclusion and results of this paper and to have an understanding of how effective is the research [26–33]. The evaluation is divided into survey and interview form [34–43], using a great user experience platform to make it easier and appealing for the participants to conduct this evaluation. We will reach people from different age gab to get an accurate variety result.

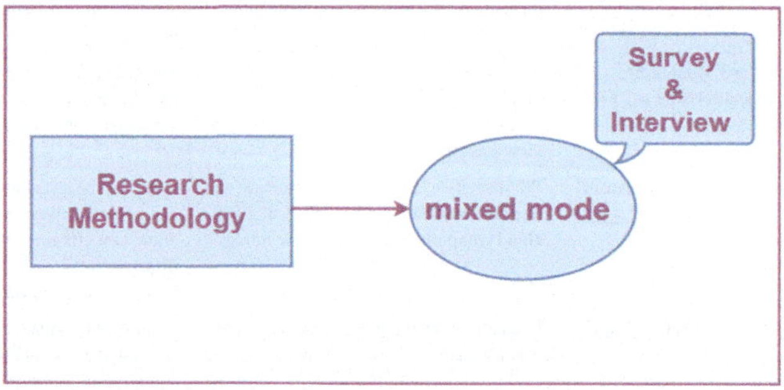

Fig. 4. Evaluation method

The convenient sampling approach refers to the area or a sample of individuals that is easy to reach and access for the sake of research [29]. The target participants were classified based on the variety of age (18–85 years old) as we focused on the independent individuals to create more accurate results in terms of the cost element, as we assumed that individuals aged bellow 18 or above 85 might not be financially independent. For instance, individuals aged bellow 18 might not have the required financial knowledge and experience to contribute to the research study, as this medication reminder device will mainly influence the cost. The survey and interview will be distributed equally to the participants however, to have more significant and accurate results for the study the equal distribution for the evaluation is grouped into:

1. The individuals will be divided into low-income families (<RM 5000 per month) and high-income families (>RM 5000 per month).
2. Health condition of the individuals, where the individuals classified into unhealthy individuals with chronic diseases or any of those who have to take medications regularly and others who are healthy and might only take health supplements. This method of categorized individuals is quite important to have an exact informative of the effectiveness.
3. Educational level of the participants, higher educational level individuals were collected from universities as academic individuals and low educational level individuals which are those participants obtained from local and public shops.

Therefore, the evaluation form will be collected mainly from educational institutions, local markets, and medical centres. To manage a detailed evaluation for the study, the educational institutions where the evaluation will take place (UCSI UNIVERSITY, and UNIVERSITY MALAYA) were selected carefully based on the variety difference of students in each university as UCSI UNIVERSITY is a private university with different students principles, on the other hand, UNIVERSITY OF MALAYA is a governmental university with different values.

Regarding the health centres, the data collection will take place in Columbia Asia hospital, and the University of Malaya medical centre, the two hospitals are chosen because of the diverse patients each hospital has, as Columbia Asia is a private multinational hospital, however, UM medical centre is a public hospital.

4 Results Findings

The results will be categorized into two main parts the survey analysis the interview analysis which is considered the detailed technical results. The research is conducted in a multi-location environment based on certain criteria of the participants to carry out a detailed result for this research. In Table 3, the full characteristic of the participants is introduced to evaluate the research methodology environment and the participants.

The participants of both hospital and university believe that the cost of hospital visitation is quite high, and the majority of the responses strongly encouraged developing the medication reminder device, however, the percentage of hospital patients are greater than the students regarding the cost-effective as shown in Fig. 4.

The effectiveness of using the device remote technique is measured from the viewpoint of both categories of participants. Most of the participants (90%) feel positive about the innovative device and selected a high level of agreement regarding the significant benefit. Furthermore, the patients had a greater percentage than the students regarding the urgent need for this device as shown in Fig. 5. On the other hand, in terms of the trust between the users and the product, it is revealed that 97% of participants agreed on setting the security element as a factor to strengthens the product trust. Figure 6 illustrates the participants' positive agreement on having a finger recognition or code security measurement to protect medications from unauthorized access. The hospital participants had a stronger belief of agreement than the educational institution participants regarding the security measurements and its effect.

Table 3. Biography of participants

Item	Parameter	Percentage
1	Age	
	18–20	19%
	21–30	35%
	31–45	22%
	45–85	24%
2	Location	
	Hospital	
	Columbia Asia	28%
	UM specialist centre	25%
	Educational institution	
	UCSI university	23%
	University of Malaya	25%
3	Income*	
	Low income	69%
	High income	31%
4	Health status	
	Vit deficiency	39%
	Chronic illness	30%
	Regular disease	31%

(*) Low income (Below RM5000); High income (Above RM5000).

Based on the survey data analysis above, most participants from both categories have published significant positive responses regarding the innovative medication device and especially its effect on their financial status, health, and device trust. The patients have reported vital needs and necessity to develop the medication reminder device, on the other hand even though the educational institution participant's analysis outcome is a significant agreement however it has a slightly lower percentage than the other group. To justify the survey results and to carry out an informative specific result an additional interview was conducted.

From the educational centre, 70% responses with average income families stated that they pay more than RM200 for their hospital visitation and they struggled with this amount of cost which they referred to as (highly expensive) for their financial situation regardless of their income, however, the other 30% of participants stated that they do not consider hospital visitation costly because they do not need going to the hospital which considers invalid for this study. On the other hand, 94% responses from patients group stated their extreme need for such a modern device for the sake of the cost reduction as

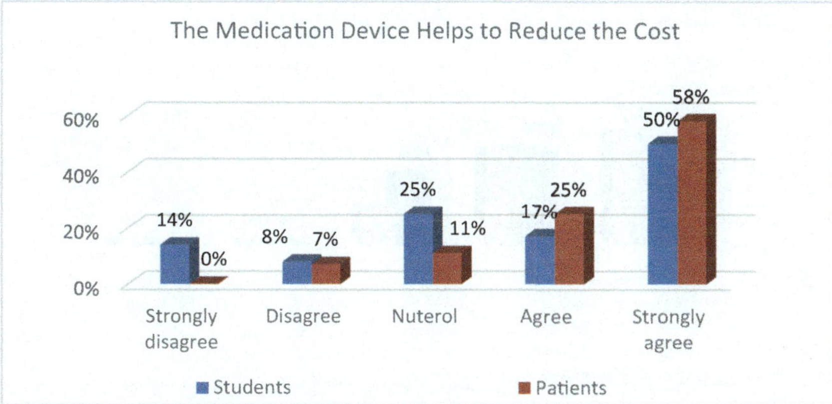

Fig. 5. Effect of cost element of the device

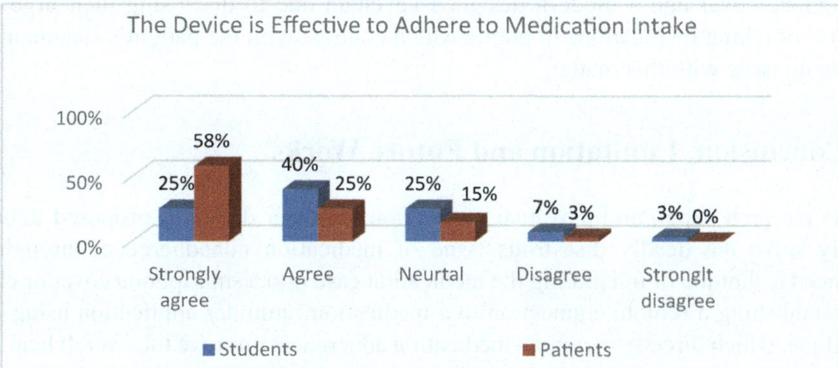

Fig. 6. Effectiveness of the medication device

they have been struggling with the doctor visitation fees related to misuse of medicine as a significant number of the patients had different chronic diseases, which constantly requires strike medication intake.

Furthermore, 95% of the interview responses from both categories insisted and significantly encouraged setting a fingerprint recognition as a security measure for the medication case as the majority stated they are always worried about child medication poison which is being a global problem with many research study to solve the issue as the traditional medicine bottle lock is not effective to prevent unauthorized access [44, 45]. This is also reflected by Fig. 7 by the high "Strongly Agreed and Agreed" statistic. The significant difference of the interview responses between the two groups indicated that the patients concentrated the accuracy and the smartphone connection technical effectiveness, however, the academic participants constantly supported the design effectiveness of the device and encouraging the viral trend of integrating medications into

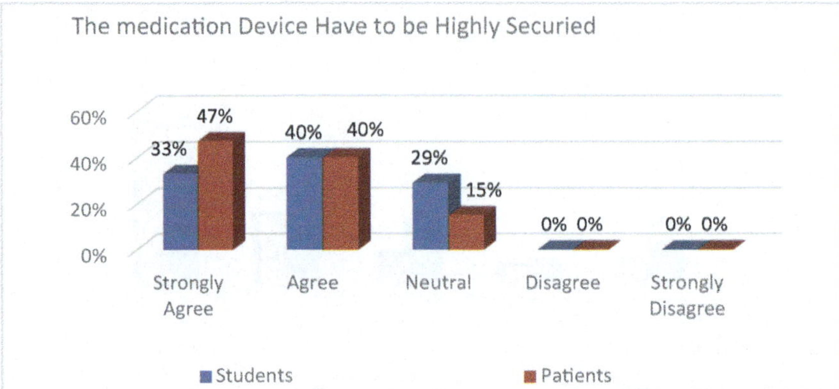

Fig. 7. Security measurements of the device

smartphone cover and a modern designed keychain due to disclosing their urge discomfort of taking medications in public which contrast with the patient's statements of having no issue with this matter.

5 Conclusion, Limitation and Future Works

In this research paper, an innovative medication reminder device is proposed to completely solve the deadly disastrous issue of medication nonadherence, through an advanced technique of integrating the medication case into a smartphone cover or chain and establishing a remote connection to a medication reminder application using IOT technique, which directly increases medication adherence, improve the overall health of users and save many lives eventually prevents a huge amount of wasted money on hospital visitations and saving the world from economic failure. The device implements a fingerprint recognition security measurement to avoid unauthorized access to the medicine, therefore preventing child poison accident.

Although this device itself could motivate the user to stick to their medication and achieve a safe and ultimate secure method for their use for medication intake and also to maintain good health and fast recovery with less spend of money as the impact can be visible from the user's responses and consistency. If many users were to participate to use this device and successfully achieved their fast recovery and good health the worldwide population can be a better and healthier nation. However, it requires more time to compare the significant difference of each person under different health issues. A future enhancement to this study is developing the device prototype and adding a feature of the medical staff interface to globalize the platform and enhance the communication between the user and the doctor.

Conflict of Interest. The authors declare that they have no conflict of interest.

References

1. Thomas, R.H., Abdul Rahim, M.I.: Gamification of medication adherence in epilepsy. Seizure Eur. J. Epilepsy **52**, 11–14 (2017)
2. Atallah, Z.A., JosephNg, P.S., Loh, Y.F.: JomMedicMinder: live or dead. In: IEEE Symposium on Computer Applications & Industrial Electronics, Kuala Lumur, Malaysia, pp. 100–105 (2021)
3. Jaya, M.K.A., Endarti, D., Kartika, G.A.A., Veryanti, P.R., Swastini, D.A.: The role of medication reminder technology as an enhancement of patients compliance. Int. J. Pharmac. Res. **12**(1), 418–427 (2020)
4. Jimenez, G., Lum, E., Huang, Z., Theng, Y.L., Car, J.: Reminders for medication adherence in type 2 diabetes management apps. J. Pharm. Pract. Res. **50**(1), 78–81 (2020)
5. Peng, Y., et al.: Effectiveness of mobile applications on medication adherence in adults with chronic diseases: a systematic review and meta-analysis. J. Manage. Care Pharm. (JMCP) **26**(4), 550–561 (2020)
6. Ahmed, N.J.: The rate of adherence to antibiotics and reasons for non-adherence among the public. J. Pharmac. Res. Int. **32**(7), 42–47 (2020)
7. Kim, B.C., Bajpai, V.K., Park, Y.H., Rather, I.A.: Self-medication and antibiotic resistance: crisis, current challenges, and prevention. Saudi J. Biol. Sci. **24**(4), 808–812 (2017)
8. Kinthada, M.R., Bodda, S., Mande, S.B.K.: eMedicare: MHealth solution for patient medication guidance and assistance. In: 2016 International Conference on Signal Processing, Communication, Power and Embedded System (SCOPES), Paralakhemundi, pp. 657–661 (2016)
9. Ramljak, M.: Smart home medication reminder system. In: 017 25th International Conference on Software, Telecommunications and Computer Networks (SoftCOM), Split, pp. 1–5 (2017)
10. Kalantarian, H., Alshurafa, N., Le, T., Sarrafzadeh, M.: Non-invasive detection of medication adherence using a digital smart necklace. In: 2015 IEEE International Conference on Pervasive Computing and Communication Workshops (PerCom Workshops), St. Louis, USA, pp. 348–353 (2015)
11. Singh, N., Varshney, U.: Medication adherence: a method for designing context-aware reminders. Int. J. Med. Inf. **132**, 103980 (2019)
12. Kassem, A., Antoun, W., Hamad, M., El-Moucary, C.: A comprehensive approach for a smart medication dispenser. Int. J. Comput. Digit. Syst. **8**(2), 131–141 (2019)
13. Ameta, D., Mudaliar, K., Patel, P.: Medication reminder and healthcare – an android application. Jo. Manag. Public Sect. Inform. Commun. Technol. **6**(2), 39–48 (2015)
14. Khedkar, S., Deshpande, S., Choudhari, M., Charles, D.: A smart pill box to remind of consumption using IoT. Int. J. Comput. Appl. **128**(1), 38–40 (2018)
15. Abdul Minaam, D.S., Abd-ELfattah, M.: Smart drugs: improving healthcare using smart pill box for medicine reminder and monitoring system. Future Comput. Inf. J. **3**(2), 443–456 (2018)
16. Arora, K., Singh, U.: Smart pill dispenser using internet of things. Int. J. Eng. Res. Technol. (IJERT) **07**(07), 486–489 (2018)
17. Pérez-Jover, V., Sala-González, M., Guilabert, M., Mira, J.: Mobile apps for increasing treatment adherence: systematic review. J. Med. Internet Res. **21**(6), e12505 (2019)
18. Fallah, M., Yasini, M.: A medication reminder mobile app: does it work for different age ranges. In: Informatics for Health: Connected Citizen-Led Wellness and Population Health, , pp. 68–72. IOS Press (2017)
19. Zanjal, S., Talmale, G.R.: Medicine reminder and monitoring system for secure health using IoT. In: International Conference on Information Security & Privacy, Nagpur, India, pp. 471–476 (2015)

20. Panda, N.K., Bhardwaj, S., Bharadwaj, H., Singhvil, R.: IOT based advanced medicine dispenser integrated with an interactive web application. In: International Conference on Advances in Mathematical Sciences, Cambridge, UK, pp. 46–48 (2018)
21. Fallahzadeh, R., Minor, B., Evangelista, L.S., Cook, D.J., Ghasemzadeh, H.: Demo abstract: mobile sensing to improve medication adherence. In: 16th ACM/IEEE International Conference on Information Processing in Sensor Networks, Pennsylvania, USA, pp. 143–145 (2017)
22. Márquez Contreras, E., et al.: Specific hypertension smartphone application to improve medication adherence in hypertension: a cluster-randomized trial. Curr. Med. Res. Opin. **35**(1), 167–173 (2018)
23. Rajanayagam, J., Bishop, J.R., Lewindon, P.J., Evans, H.M.: Paracetamol-associated acute liver failure in Australian and New Zealand children: high rate of medication errors. J. Drug Issues **100**(1), 77–80 (2015)
24. Choudhry, N.K., Krumme, A.A., Ercole, P.M.: Effect of reminder devices on medication adherence: the REMIND randomized clinical trial. JAMA Intern. Med. **177**(5), 624–631 (2017)
25. Huh, J.-H., Seo, K.: Blockchain-based mobile fingerprint verification and automatic log-in platform for future computing. J. Supercomput. **75**(4), 3123–3139 (2018). https://doi.org/10.1007/s11227-018-2496-1
26. Hox, J., De Leeuw, E., Zijlmans, E.: Measurement equivalence in mixed-mode surveys. Front. Psychol. **6**(1), Article 87 (2015)
27. Blanchflower, T., Leavy, P.: Research Design: Quantitative, Qualitative, Mixed Methods, Arts-Based, and Community-Based Participatory Research Approaches. The Guilford Press, New York (2017). ISBN 9781462514380. 300 p. (Paperback). Family Consum. Sci. Res. J. **47**(1), 101–102 (2018)
28. JosephNg, P.S., Eaw, H.C.: Making financial sense from EaaS for MSE during economic uncertainty. Springer Adv. Inf. Commun. **1**(2), 976–989 (2021)
29. JosephNg, P.S., Loh, Y.F., Eaw, H.C.: Grid computing for MSE during volatile economy. In: International Conference on Control, Automation and Systems, pp. 709–714. IEEE Explore, Busan (2020)
30. JosephNg, P.S., et al.: EaaS: available yet hidden infrastructure inside MSE. In: 5th International Conference on Network, Communication and Computing, ACM International Conference Proceeding Series, Kyoto, Japan, pp. 17–20 (2016)
31. JosephNg, P.S., Kang, C.M., Choo, P.Y., Wong, S.W., Phan, K.Y., Lim, E.H.: Beyond cloud infrastructure services in medium size manufacturing. In: International Symposium on Mathematical Sciences & Computing Research, pp. 150–155. IEEE Explore, Ipoh (2015)
32. Joseph Ng P.S., Choo, P.Y., Wong, S.W., Phan, K.Y., Lim, E.H.: Malaysia SME ICT During Economic Turbulence. In: International Conference on Information & Computer Network, Singapore, pp. 67–71 (2012)
33. Joseph Ng, P.S., Yin, C.P., Wan, W.S., Nazmudeen, M.S.H.: Energizing ICT infrastructure for Malaysia SME during economic turbulence. In: Student Conference on Research and Development, pp. 328–322. IEEE Explore, Cyberjaya, Malaysia (2011)
34. JosephNg, P.S.: Economic turbulence & EaaS grid computing. Int. J. Bus. Forecast. Mark. Intell. (2021, forthcoming)
35. JosephNg, P.S., Eaw, H.C.: Still technology acceptance model? Reborn: exostructure as a service model. Int. J. Bus. Inf. Syst. (2021, forthcoming)
36. JosephNg, P.S.: EaaS infrastructure disruptor for MSE. Int. J. Bus. Inf. Syst. **30**(3), 373–385 (2019)
37. JosephNg, P.S.: EaaS optimization: available yet hidden information technology infrastructure inside medium size enterprises. J. Technol. Forecast. Soc. Change **132**(July), 165–173 (2018)

38. JosephNg, P.S., et al.: Exostructure services for infrastructure resources optimization. J. Telecommun. Electron. Comput. Eng. **8**(4), 65–69 (2016)
39. JosephNg, P.S., Kang, C.M.: Beyond barebone cloud infrastructure services: stumbling competitiveness during economic turbulence. J. Sci. Technol. **24**(1), 101–121 (2016)
40. Joseph, N.P.S., Mahmood, A.K., Choo, P.Y., Wong, S.W., Phan, K.Y., Lim, E.H.: Barebone cloud IaaS: revitalization disruptive technology. Int. J. Bus. Inform. Syst. **18**(1), 107–126 (2015)
41. Joseph, N.P.S., Mahmood, A.K., Choo, P.Y., Wong, S.W., Phan, K.Y., Lim, E.H.: IaaS cloud optimization during economic turbulence for malaysia small and medium enterprise. Int. J. Bus. Inform. Syst. **16**(2), 196–208 (2014)
42. Joseph, N.P.S., Mahmood, A.K., Choo, P.Y., Wong, S.W., Phan, K.Y., Lim, E.H.: Battles in volatile information and communication technology landscape: the Malaysia small and medium enterprise case. Int. J. Bus. Inform. Syst. **13**(2), 217–234 (2013)
43. Joseph Ng, P.S., Choo, P.Y., Wong, S.W., Phan, K.Y., Lim, E.H.: Hibernating ICT infrastructure during rainy days. J. Emerg. Trends Comput. Inform. Sci. **3**(1), 112–116 (2012)
44. Martínez-Mesa, J., González-Chica, D.A., Duquia, R.P., Bonamigo, R.R., Bastos, J.L.: Sampling: how to select participants in my research study? Braz. Ann. Dermatol. J. **91**(3), 326–330 (2016)
45. Jovanov, B.D.E., Talukder, B., Schwebel, D.C., Douglas Evans, W: Design and feasibility of a safe pill bottle. Appl. Syst. Innov. **1**(2), Article 13 (2018)

Cross-Train: Machine Learning Assisted QoT-Estimation in Un-used Optical Networks

Ihtesham Khan(✉), Muhammad Bilal, and Vittorio Curri

DET, Politecnico di Torino, Torino, Italy
ihtesham.khan@polito.it

Abstract. The quality of transmission (QoT) estimation of lightpaths (LPs) has both technological and economic significance from the operator's perspective. Typically, the network administrator configures the network element (NE) working point according to the specified nominal values given by vendors. These operational NEs experienced some variation from the given nominal working point and thus put up uncertainty during their operation, resulting in the introduction of uncertainty in estimating LP QoT. Consequently, a substantial margin is required to avoid any network outage. In this context, to reduce the required margin provisioning, a machine learning (ML) based framework is proposed which is cross-trained using the information retrieved from the fully operational network and utilized to support the QoT estimation unit of an *un-used* sister network.

Keywords: Machine learning · QoT-estimation · Generalized SNR

1 Introduction

The latest evolving technologies such as 5G, virtual and augmented reality, internet of things (IOTs), and different cloud platforms increase the trend of the global internet traffic [1]. This upsurge of the latest technologies and bandwidth-hungry applications has put on high demand and new requirements for capacity improvement and optical networks' reliability. To accommodate this remarkable growth of internet traffic and maximize profits on *CAPEX* assets, the network operator always requests the full exploitation of the remaining capacity of the already deployed infrastructure. To this aim, the data transport layer must be pushed to reach the maximum capacity limit. The primary technologies for exploiting data transport are the dense-wavelength-division multiplexed (DWDM) together with coherent transmission. These technologies pave a path for evolving technologies like elastic optical networks (EONs) and optical software-defined networking (Optical-SDN). The EONs enable efficient utilization of the available spectrum by enabling the network controller to offer flexible assignment of spectrum slices to the particular traffic request [2]. At the same time, the SDN controller empowers the separate configuration of the working points of each NE and provides the mean for a customized network management system.

The foundation step towards elastic and customized network management is the abstraction of the optical transport network as a topological grid subjective with the

M. Ma (Ed.): ICTCE 2020, LNEE 797, pp. 78–87, 2022.
https://doi.org/10.1007/978-981-16-5692-7_9

GSNR degradation during the propagation through optical line systems (OLSs) which comprise fiber spans followed by amplifies [3]. Typically, OLSs are managed with the centralized operating system in the control plane [4]. This centralized controller adjusts the amplifier operations and subsequently defines the QoT deprivation. The exploitation of the exact nominal working point enables well estimation of the total LP QoT. Hence, during the provisioning of LP, a smaller system margin is demanded, and subsequently, more traffic can be accommodated, assisting improved utilization of the mounted infrastructure.

In the current frame of reference, QoT is characterized in terms of the generalized SNR (GSNR), which incorporates the impact of ASE noise and NLI accumulation [5]. The flexible transceiver considered in this work is characterized by providing an OSNR threshold for a given modulation format; the existing GSNR of a given LP describes the path viability. Thus, the main application of SDN towards the transport layer is a QoT estimator (QoT-E); providing the network information, the QoT-E engine calculates the GSNR over a particular LP. Referring to the Telecom Infra Project [5, 6], it has been widely validated by providing the precise information of the physical layer; a QoT-E engine can deliver a precise estimation of GSNR. Generally, NEs suffered from a variation in the working point due to the hardware (HW) aging, the variation of spectral load at OLS, and different environmental effects during field operations. These induced variations affect the actual GSNR estimation using the nominal values by the QoT-E engine [7]. Additionally, amplifiers' ripples gain, noise figure, and the fiber connector/insertion losses also yield GSNR uncertainties. Consequently, the calculated nominal GSNR on a given LP demands a reasonable margin deployment to prevent any network outage [8].

This work's primary motivation is to reduce the GSNR uncertainty of a particular LP and, consequently, facilitate reliable path calculation for the LP provision at the lowest possible margin. The proposed work is simulated considering an open optical network setup, where the network controller deploys the QoT-E engine as an application program interface (API). Suppose the controller is provided with the exact knowledge network condition, i.e., an *accurate characterization* of operating parameters of every NE. In this case, the QoT-E can calculate the GSNR with reasonable precision, as demonstrated in [5, 6]. In contrast to this, during the unavailability of the *actual characterization* of the operating point of every NE, the QoT-E depends on the *nominal characterization* of the working point of every NE. The QoT-E engine exploits this nominal characterization and calculates a nominal GSNR. The obtained nominal GSNR has an uncertainty in its measurement as formerly described.

In the current investigation, the information retrieved from an in-service operating network is used to cross-training the ML framework operating in the controller of another *un-used* sister network. This cross-trained module supports the QoT-E unit of an *un-used* network in estimating accurate GSNR of LP. The proposed work considers two different networks based on topology, but both are the same on the install HW, like fiber class and erbium-doped fiber amplifiers (EDFAs). The networking topologies considered in this analysis; the European Union (EU) network topology as a well-operative network and USA network topology as *un-used* sister network (see Fig. 1a and b). The dataset is obtained synthetically by perturbing the estimated QoT of LP with proper random

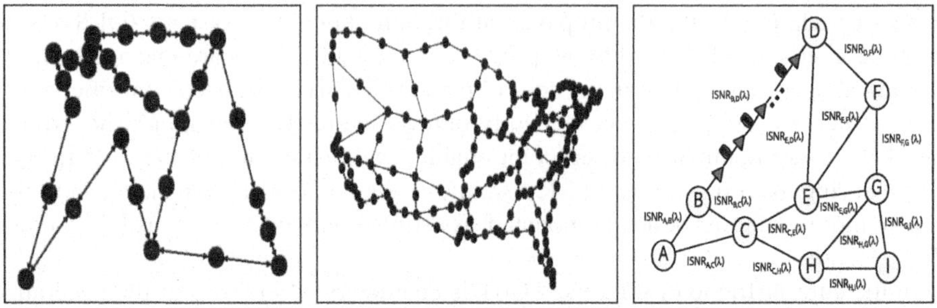

(a) (b) (c)

Fig. 1. (a) EU network topology (b) USA network topology (c) Abstraction of optical network

spectral-load-dependent NE parameters; specifically, we focused on EDFA ripples and uncertainties in connector losses.

The ML framework proposed in the current simulation scenario reduces the GSNR uncertainty of the LP of an *un-used* sister network. The presented ML module corrects the GSNR calculation of the QoT-E unit of an *un-used* sister network, which exploits nominal NE parameters to estimate it. The data-driven techniques based on ML are already used in the optical networks for different applications; in [9, 10] the authors proposed an ML-based technique for network performance monitoring. In [11, 12] ML framework is proposed for QoT estimation. More than a few data-driven procedures for QoT estimation of LP prior to its real deployment in the network are demonstrated in [13–15]. In [16], the domain adaptation method is used to estimate the QoT of LP. The authors in [17] achieved the precision in QoT estimation using active learning and domain adaptation procedures. Finally, a comprehensive review of ML-employed applications in optical networks is reported in [18].

The core distinction of this work from the already performed investigations is that this scheme proposed the cross-training technique to train an ML module efficiently. Besides this, the cross-trained ML module operates synergically with the QoT-E engine in the network controller. This synergic use's primary purpose is to reduce the GSNR uncertainties induced by EDFA gain ripples, noise figure, and the uncertainties induced by fiber connector losses.

2 Networks Model and Data Generation

In the present work, a core optical network is mapped as a topological graph having edges represented as OLSs. In contrast, nodes are portrayed as a site of reconfigurable optical add-drop multiplexing (ROADM). The considered OLSs include a span of fibers separated by equidistant amplifiers. The OLSs are managed by a centralized controller and are supposed to be operated at optimal operating point [19]. The controller responsible for configuring the OLS exploits the parameters related to the physical layer. Regarding these parameters, the more vulnerable parameters are the fiber connector/insertion losses, amplifier ripples gain (Uniform Variation of 1 dB), and noise figure ([6–11] dB), typically varying with the spectral-load. Besides this, fiber losses such as fiber attenuation ($\alpha = 0.2$ dB/km) and dispersion ($D = 16.0$ ps/nm/km) also take into account to get a realistic simulating environment. The statistics of considered connector losses are defined by an exponential distribution ($\lambda = 4$) termed in the analysis [20]. The metric of QoT, i.e., the $GSNR$ of any given LP propagated from source node towards destination node traversing through a definite OLSs connected them is presented as $1/GSNR = \sum_n 1/GSNR_n$, where n is the total number of OLSs connecting the given source node to the destination node revealed in Fig. 1c. The $GSNR$ metric of the given LP is presented by:

$$\mathrm{GSNR} = \frac{P_{\mathrm{RX}}}{P_{\mathrm{ASE}} + P_{\mathrm{NLI}}} = \left(\mathrm{OSNR}^{-1} + \mathrm{SNR}_{\mathrm{NL}}^{-1}\right)^{-1}, \tag{1}$$

where $\mathrm{OSNR} = P_{\mathrm{Rx}}/P_{\mathrm{ASE}}$ is the optical signal to noise ratio detectable by optical channel monitors, $\mathrm{SNR}_{\mathrm{NL}} = P_{\mathrm{Rx}}/P_{\mathrm{NLI}}$ is the nonlinear SNR induced by NLI only and is recovered using the digital signal processing constellation. P_{Rx} is the channel power at the receiver end, P_{ASE} is the ASE noise power, and $\mathrm{P}_{\mathrm{NLI}}$ is the accumulated NLI power. The GSNR is associated with the bit error rate (BER) by the BER vs. OSNR description for the particular modulation format [5].

To limit the computational effort, the considered OLSs operate no more than 76 channels around the basic 50 GHz grid on the C-band, having entire bandwidth of almost 4 THz. Indicating standard 96 channels on the entire C-band does not anticipate significant differences in the results. The considered transceivers work at 32 GBaud, and the configured EDFAs work at a fixed output power mode supplying 0 dBm power against each channel. The simulated network connections are supposed to be operated using standard single-mode fiber (SSMF) with a maximum span-length of 80 km. An open-source network simulation tool called GNPy is used to mimic the real field data to obtain a realistic dataset. Moreover, the considered tool is selected as it is more reliable and well-tested (see [21, 22]). Usually, this library creates the network templates for the physical layer by simulating an end-to-end environment [23]. The open-source GNPy library is resolved spectrally and is instituted on the generalized-Gaussian-noise (GGN) model [24]. Exploiting the spectral resolution capability, the GNPy is constituted to clone the network data in the real field environment. The cloned dataset includes channel power at receiver, ASE noise, NLI accumulation, the GSNR for every LP, and finally, the total spans traversed from source-to-destination (s $\rightarrow$ d). Considering the optimal signal power, the ASE noise is the main factor as at optimal level ASE is always duple the NLI [3, 25]. Unusually, ASE is also very tricky to estimate, as it varies with the operating condition of EDFAs [26], which ultimately hinges on the spectral-load [27]. In this perspective, the engendered dataset is perturbed by changing the highly fragile characteristics of EDFA, typically amplifier ripple gain and noise figure.

The mimicked dataset comprises two separate datasets; one section describes an *in-service* network, whereas the other denotes an *un-used* network. The considered networks are characterized by distinct topologies having similar fiber class and transmission apparatus. Nevertheless, they are exclusive in amplifier parameters like amplifier ripple gain, noise figure, and fiber connector losses. Following the network composition of both datasets (*in-service & un-used*), the next section is the spectral-load. In the present simulation, the spectral-load of a given simulated link is the subset of 2^{76} possible realization of spectral-load against considered 76 channels. In the reflected subset of spectral-load, each (s$\rightarrow$d) pair has 1024 combinations of random traffic having maximum occupation

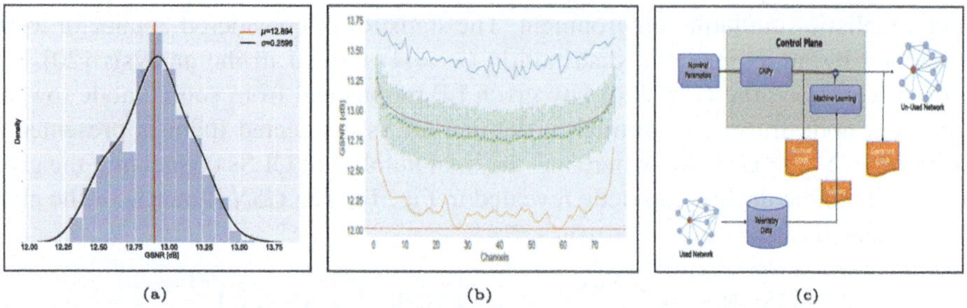

Fig. 2. (a) GSNR distribution (b) Train path GSNR statistical analysis (c) Model orchestration

ranges between 34% to 100% of overall considered bandwidth. The initial subset of the dataset is created against the EU network topology (*in-service network*). In contrast, the other subset of the dataset is created against the USA network topology (*un-used* network).

3 GSNR Statistical Analysis

In the current frame of reference, to compute the GSNR of *un-used* network, the network controller of this nascent network can only depend on the nominal explanation of the system (noise figure = 8.75 dB, gain ripple = flat ripple and insertion losses = 0.3 dB). Exploitation exclusively this nominal behavior of a network, the network controller calculates GSNR, a nominal one. This estimated GSNR has a certain level of ambiguity anticipated by the discrepancy in NEs operating points. Figure 2a indicates the GSNR distribution for the entire WDM comb for the given path *Vienna* → *Warsaw*. Observing Fig. 2a, it is well demonstrated that the given GSNR for path *Vienna* → *Warsaw* is distributed across the mean value, observing the probability density function it is well approximated as Gaussian. Similar behavior is observed for the other simulated links of *un-used* network. Going into more details, Fig. 2b reveals similar outcomes for all wavelengths on the same *Vienna* → *Warsaw* path. In Fig. 2b, the statistical breakdown of a particular variation is depicted. Besides this assessment, a small number of significant concerns evolve by calculating the average of the GSNRs for full train realizations of the given path, introduced in Fig. 2b. The average of GSNRs presented by green dots characterizes the OLSs module, ranging between 12.75 dB, and 13.27 dB, with standard deviations (error bars) 0.20 dB to 0.28 dB. A purple line depicts the nominal GSNRs for the given path. The bluish and orange lines bordered the maximum and the minimum GSNRs values for every channel. The dotted reddish line specifies the global minimum GSNR of 12.02 dB for the given path. The current GSNR indicators show up two methods to deliver QoT.

In the first approach, reflecting a worst-case (WC) setup without any knowledge, the constant GSNR threshold should be applied for all the channels with a value smaller than an anticipated global minimum (12.02 dB) to avoid any network outage. In this approach, the fluctuation in GSNR values ranges between 13.68 dB to 12.02 dB; this creates almost a 1.6 dB of margin requested by the GSNR uncertainty.

In the second approach, we considered the nominal GSNR estimation of the core QoT-E engine. In this method, we have two states of GSNR description around the nominal one. The first one comprises those channels having GSNR values higher than the nominal GSNR estimation. The second one is those channels having GSNR values lower than the nominal estimation. To measure the uncertainty in GSNR estimation in this approach, we calculate the difference between the nominal and actual one using Eq. 2. Considering the first case, the

$$\Delta \mathrm{GSNR} = \mathrm{GSNR}_{\mathrm{nominal}} - \mathrm{GSNR}_{\mathrm{actual}}, \tag{2}$$

one with higher GSNR values than the nominal estimation reporting a maximum GSNR uncertainty of 0.85dB (the maximum difference between purple line & blue line) having negative $\Delta GSNR^{-}$ description. This $\Delta GSNR^{-}$ case is not critical as the available GSNR threshold of these channels at a transceiver is higher than the estimated nominal GSNR. The transceiver, in this case, can quickly deploy a reliable LP. In contrast to this, the second case, the one having lower GSNR values than the nominal estimation unfolding a maximum GSNR uncertainty of almost 1.25 dB (the maximum difference between purple line & orange line) having positive $\Delta GSNR^{+}$ description. Unlike the first case, the $\Delta GSNR^{+}$ case is much more critical as the available GSNR threshold of these channels at a transceiver is lower than the estimated nominal GSNR. The transceiver, in this case, will be configured with a high margin to deploy an LP reliably and keep the network in-service state.

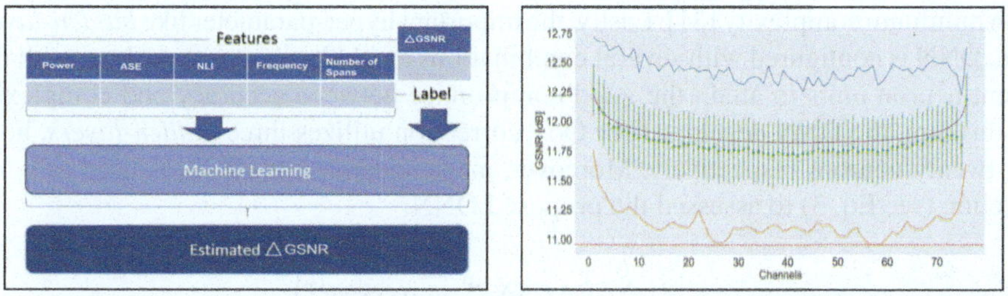

(a) (b)

Fig. 3. (a) Machine learning module (b) Test path GSNR statistical analysis

Consequently, the main challenge in the current simulating environment is dealing with a more critical $\Delta GSNR^{+}$ case. In the current study, the main objective is to decrease the error ($\Delta GSNR^{+}$) in the estimation of the QoT-E engine of the GNPy unit, in the absence of exact system parameters. To this aim, we consider the exploitation of the data retrieved from the EU network to cross-train ML unit operating on the controller of the USA network. The proposed cross-trained ML element is utilized to assist the main QoT-E engine of *un-used* USA network shown in Fig. 2c. The proposed scheme delivers a QoT rectifying apparatus that can provide precise QoT estimation of LP prior to its actual provisioning in the network.

4 Visual Inspection of Machine Learning Module

The present simulation scenario employs the data-driven ML model, which is initially cross-trained. It is then deployed in the controller of an *un-used* network, where it helps to correct the *GSNR* estimation of the core QoT-E engine. Similar to other data-driven methods, the proposed ML prototype training and testing procedures need to define the features and response variable, indicating the structural inputs and outputs. The operated well-defined features incorporate channel power at receiver, NLI accumulation, ASE noise, frequency of the channel, and the fiber spans between the given s → d node. Along with this, the manipulated response variable is the GSNR correction parameter; $\Delta GSNR$ of the given LP depicted in Fig. 3a. The overall sum of the participation features comprises 305 definitions; $(1+(4 \times 76) = 305)$ the number spans plus the channel power at receiver, the NLI accumulation, the ASE noise, and channel frequency.

The proposed ML unit obtained the perceptive ability to provide the *GSNR* correcting metric by mapping the features and response variable of an in-service network. The defined functionality is achieved by using Deep Neural Network (DNN) [28], which is an excellent data-driven model to discover the correlation among the given features and required response variable. The presented DNN in this work is structured by utilizing open-source APIs of TensorFlow© library [29], and is configured by numerous set of hyper-parameters like *training steps* = 1000, supplied by default Keras optimizer as *adaptive gradient algorithm* along with *learning rate* = 0.01 coupled with regularization $L_1 = 0.001$ [30]. Additionally, Relu is nominated as an activation function to allows the efficient interpretation of the provided input features into the desired response variable with minimum complexity [31]. Lastly, the important hyper-parameter like *hidden-layers* size, DNN is configured with several combinations of *hidden-layers* size along with different neuron units to attain the good compromise between accuracy and complexity. To this aim, the DNN developed for QoT correction utilizes three *hidden-layers*, holding twenty neurons respectively. Moreover, mean square error (*MSE*) is used as a loss function (see Eq. 3) to assessed the proposed DNN,

$$\mathrm{MSE} = \frac{\sum_{j=0}^{N} \left(\Delta GSNR_j^{\mathrm{p}} - \Delta GSNR_j^{\mathrm{a}}\right)^2}{N}, \tag{3}$$

where $\Delta\mathrm{GSNR}_j^{\mathrm{r}}$ and $\Delta\mathrm{GSNR}_j^{\mathrm{p}}$ are the actual and DNN generated predicted measurements of error in GSNR estimation of a channel under test for the jth spectral-load, correspondingly, and N is the over-all sum of ΔGSNR combination of the test set. The proposed DNN is further shaped for training, authentication, and examination by the traditional regulation 70/15/15 and is crossed-trained by ΔGSNR responses of a random spectral-load of an already deployed EU network. After proper cross-training, the cross-trained ML module is used synergically with the QoT-E agent of GNPy to assist the QoT-E agent in precisely estimating LP GSNR value before its deployment. In the current setup, the training set data entails 4096 combinations for four s → d paths (1024 realizations for each s → d pair) *Frankfurt* → *Istanbul*, *London* → *Madrid*, *Brussels* → *Bucharest*, *Vienna* → *Warsaw* of an EU network having span length of 34, 19, 30 and 7 respectively. The test set in the proposed frame work contains 6144 combinations for six s → d paths *Charlotte → Chicago, Louisville → Memphis, Memphis → Miami, Kansas City → Las-Vegas, Little Rock → Long Island, Los Angeles → Louisville* of the USA network having span length 20, 7, 24, 30, 26, 46 respectively.

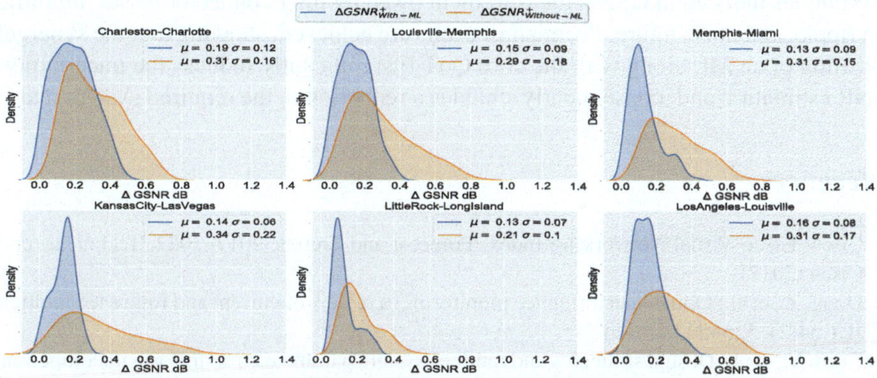

Fig. 4. Distribution of $\Delta GSNR^+_{with-ML}$ and $\Delta GSNR^+_{without-ML}$

5 Results and Discussion

The performance of the proposed DNN to reduce the error in the GSN R estimation of the QoT-E engine of an *un-used* network is summarized in this section. The proposed work considers the $\Delta GSNR^+$ case only, as this is a more critical margin as compared to ΔGSN R− described in Sect. 3.

To simplify the description of the acquired outcomes, we initially describe the results related to one Louisville → Memphis path of the USA network. Specifying the statistical analysis of GSNR of this path, we put the base reference, $\Delta GSN\,R^+$ retrieved by mirroring minimum GSNR measurement (10.81) dB demonstrated in Fig. 3b. This particular case represents a rough estimation, and it offered a reference level. This approach creates a margin of $\Delta GSN\,R^+_{without-ML} = GSN\,R_{nominal} - \Delta GSN\,R_{Globalminimum} = 1.1$ dB on the WC scenario. Next the QoT-E unit is assisted by the cross-trained DNN, the given framework delivers a definitive QoT-E allowing reduction in the margin of $\Delta GSN\,R^+_{with-ML} = GSN\,R_{nominal} - \Delta GSN\,R_{predicted} = 0.6$ dB on the similar path. The results are illustrated in Fig. 4. The reliability and scalability of the proposed scheme are further verified on five additional paths of the USA network defined in Sect. 4.

The results related to all the studied paths are shown in Fig. 4 which reported the error distribution with and without ML assistance. In Fig. 4, it is seen that the proposed DNN unit dramatically decrease the error in QoT estimation.

6 Conclusion

The proposed work exploits a data-driven method to assist in predicting the QoT of a given LP. The proposed scheme is cross-trained by using the dataset of in-service network and utilize it to decrease the GSNR margin of an *un-used* network. The core DNN unit of the cross-trained framework is developed by utilizing the TensorFlow© platform. The generated dataset is obtained synthetically for the two considered networking topologies utilizing the open-source GNPy library. The generated dataset explicitly considered the uncertainties induced in GSNR measurement owed by fiber connector losses, amplifiers gain ripple, and noise figure. Promising results are achieved, showing that the synergetic utilization of an ML along with the core QoT-E significantly reduces the uncertainty in GSNR estimation and, consequently, enables a reduction in the required system margin.

References

1. Cisco: Cisco Visual Networking Index: Forecast and Trends, 2017–2022. Technical report, Cisco (2017)
2. Dong, Z., et al.: Optical performance monitoring: a review of current and future technologies. JLT **34**(2), 525–543 (2016)
3. Curri, V., et al.: Design strategies and merit of system parameters for uniform uncompensated links supporting nyquist-WDM transmission. JLT **33**(18), 3921–3932 (2015)
4. Pastorelli, R.: Network optimization strategies and control plane impacts. In: OFC. OSA (2015)
5. Filer, M., et al.: Multi-vendor experimental validation of an open source QoT estimator for optical networks. JLT **36**(15), 3073–3082 (2018)
6. Ferrari, A., et al.: GNPy: an open source application for physical layer aware open optical networks. JOCN **12**(6), C31–C40 (2020)
7. D'Amico, A., et al.: Using machine learning in an open optical line system controller. JOCN **12**(6), C1–C11 (2020)
8. Pointurier, Y.: Design of low-margin optical networks. JOCN **9**(1), A9–A17 (2017)
9. Khan, F., et al.: Optical performance monitoring in fiber-optic networks enabled by machine learning techniques. In: 2018 (OFC), pp. 1–3. IEEE (2018)
10. Barletta, L., et al.: Qot estimation for unestablished lighpaths using machine learning. In: OFC, p. Th1J–1. OSA (2017)
11. Sartzetakis, I., et al.: Accurate quality of transmission estimation with machine learning. JOCN **11**(3), 140–150 (2019)
12. Mo, W., et al.: Ann-based transfer learning for QoT prediction in real-time mixed line-rate systems. In: 2018 OFC, pp. 1–3. IEEE (2018)
13. Khan, I., et al.: QoT estimation for light-path provisioning in un-seen optical networks using machine learning. In: ICTON, pp. 1–4 (2020)
14. Khan, I., et al.: Advanced formulation of QoT-estimation for un-established lightpaths using cross-train machine learning methods. In: ICTON, pp. 1–4 (2020)
15. Khan, I., et al.: Assessment of cross-train machine learning techniques for QoT-estimation in agnostic optical networks. OSA Continuum **3**(10), 2690–2706 (2020)
16. Di Marino, R., et al.: Assessment of domain adaptation approaches for QoT estimation in optical networks. In: 2020 OFC, pp. 1–3. IEEE (2020)
17. Azzimonti, D., et al.: Active vs transfer learning approaches for QoT estimation with small training datasets. In: OFC, pp. M4E–1. OSA (2020)
18. Mata, J., et al.: Artificial intelligence (AI) methods in optical networks: a comprehensive survey. OSN **28**, 43–57 (2018)
19. Curri, V., et al.: Design strategies and merit of system parameters for uniform uncompensated links supporting Nyquist-WDM transmission. JLT **33**(18), 3921–3932 (2015)

20. Ando, Y.: Statistical analysis of insertion-loss improvement for optical connectors using the orientation method for fiber-core offset. IEEE Photon. Technol. Lett. **3**(10), 939–941 (1991)
21. Telecominfraproject. Telecominfraproject/oopt-gnpy, September 2019
22. Ferrari, A., et al.: GNPy: an open source application for physical layer aware open optical networks. JOCN **12**(6), C31–C40 (2020)
23. Grammel, G., et al.: Physical simulation environment of the telecommunications infrastructure project (TIP). In: OFC, p. M1D–3. OSA (2018)
24. Cantono, M., et al.: On the interplay of nonlinear interference generation with stimulated raman scattering for QoT estimation. JLT, **PP**(99) 1 (2018)
25. Ferrari, A., et al.: Observing the generalized SNR statistics induced by gain/loss uncertainties. In: 2019 ECOC. IEEE (2019)
26. Brian, T., et al.: Towards a route planning tool for open optical networks in the telecom infrastructure project. In: OFC/NFOEC 2018 (2018)
27. Bolshtyansky, M.: Spectral hole burning in erbium-doped fiber amplifiers. JLT **21**(4), 1032–1038 (2003)
28. Bishop, C.M.: Pattern Recognition and Machine Learning. Springer, Heidelberg (2006)
29. Abadi, M., et al.: Tensorflow: a system for large-scale machine learning. In: 12th {USENIX} ({OSDI} 16), pp. 265–283 (2016)
30. Duchi, J., et al.: Adaptive subgradient methods for online learning and stochastic optimization. JOMLR **12**(Jul), 2121–2159 (2011)
31. Nwankpa, C., et al.: Activation functions: comparison of trends in practice and research for deep learning (2018)

Deep Reinforcement Learning Based Routing Scheduling Scheme for Joint Optimization of Energy Consumption and Network Throughput

Binbin Ye[1], Wei Luo[1], Ruikun Wang[2], Zhiqun Gu[2](✉), and Rentao Gu[2](✉)

[1] Liare NARI Group Corporation/State Grid Electric Power Research Institute, Nanjing Nari Information & Communication Technology Co., Ltd., Nanjing, China

[2] State Key Laboratory of Information Photonics and Optical Communication, Beijing University of Posts and Telecommunications, Beijing 100876, China

{guzhiqun,rentaogu}@bupt.edu.cn

Abstract. With traffic demands growing exponentially, a great number of new network applications emerging, traffic load balancing and resource utilization have become the key issues that severely affect network performance of data center networks (DCNs). The efficient routing scheduling scheme is considered as the key factors to affect the network throughput and resource consumption. To maximize the network throughput and reduce the resource consumption, this paper investigates the efficient routing scheduling scheme by joint optimizing the network throughput and energy consumption. We first formulated the problem as a mixed integer nonlinear program (MINLP) problem, and then introduce the deep reinforcement learning based routing scheduling problem to solve it. Results show that the proposed scheme can significantly improve the network performance.

Keywords: Data center networks (DCNs) · Routing scheduling · Mixed integer nonlinear program (MINLP) problem · Deep reinforcement learning

1 Introduction

With traffic demands growing exponentially and larger number of new network applications emerging, the traffic load and resource utilization have become the key factors that affect network performance of data center networks (DCNs) [1]. Especially, designing the efficient routing scheduling strategies with the optimal network throughput and resource utilization becomes a key issue for DCNs.

There are lots of routing schemes, which establish a routing table, and then run routing protocols to exchange routing information with each other to share the routes [2]. Note that, one path will be chosen to forward more than two traffic demands because of the same destination nodes, which leads to network congestion. To reduce the network congestion, Pointurier *et al.* [3] proposed a QoS routing algorithm based on Dijkstra algorithm to select routes that meet bandwidth constraints for traffic flow, which tries

M. Ma (Ed.): ICTCE 2020, LNEE 797, pp. 88–97, 2022.
https://doi.org/10.1007/978-981-16-5692-7_10

to ensure the network QoS. In these studies, the shortest paths from the source nodes to the destination nodes are calculated, which is inefficient and not suitable for large-scale network environment. Software defined network (SDN) is proposed to separate the complex logic control involved in packet forwarding from switches/routers and other devices, it controls the network by software programming, so as to achieve the purpose of freely controlling traffic in different network scales. Then, a multi-path routing algorithm based on SDN load balancing is proposed [4], which calculates all the paths that can forward the flow, the path with the least link load is selected. However, with the expansion of network scale and the rapid growth of the number of network applications, the current network becomes highly dynamic and complicated. All the above are model-based routing algorithms, which have challenges in model hypothesis and establishment, especially for the complex network environment. In addition, these algorithms have poor scalability and universality in different scenarios.

Due to the rapid development of artificial intelligence (AI), machine learning (ML) has made very great progress and being widely concerned by the academic community [5]. Also, ML has been considered as a promising technology to solve routing optimization problems [6]. On the one hand, ML can quickly obtain the routing close to the optimal solution through training; on the other hand, ML does not need an accurate mathematical model of the underlying network [7]. However, DCNs are complex and dynamic system, the routing algorithm based on ML needs continuous learning and training for different network scenarios, which will greatly increase the network overhead [8]. To solve this problem, we consider using reinforcement learning (RL) for intelligent control. Due to RL needs to maintain a lot of information such as network status, action, reward information, and so on, that occupies a lot of storage space. Therefore, this paper combines deep learning (DL) and RL, using neural network instead of Q table learning in the process of RL process, and then adopts deep reinforcement learning (DRL) to achieve intelligent routing scheduling, and selects the best data transmission path to improve network performance.

This paper studies the routing scheduling problem for DCNs, and proposes a deep reinforcement learning based two objectives optimization routing scheduling (PEARL) scheme. PEARL first formulates the data flow routing problem as a mixed integer nonlinear programming (MINLP) with two objectives, i.e., maximizing the network throughput and minimizing the energy consumption. Here, the energy consumption is defined as sum of energy of all the links. Then, DRL is utilized to solve the routing scheduling problem, and the simulation results demonstrate the performance of PEARL.

2 Network Model and Problem Formulation

We consider the data center networks (DCNs) as an undirected graph G, where $G = (V, E)$, and V is denoted as the set of switches, E is the set of links. The traffic demands are given, and the set is denoted as M. For each demand $m, f_{s,d}$ indicates the data flows from the origin switch s to the destination switch d, and data rate is denoted as q.

2.1 Link Model

For each link s, x_e is defined as the sum of traffics flows on e ($e \in E$), i.e.,

$$x_e = \sum_s \sum_d f_{s,d}^e, \forall s, d \in V, s \neq d. \tag{1}$$

The energy consumption of link e is calculated as follows [1],

$$g(x_e) = \sigma + \mu x_e^\alpha, 0 \leq x_e \leq \beta C_e, \tag{2}$$

where σ, μ, and α are both constants. σ is the energy consumption when the link is idle, C_e is the bandwidth capacity of the link, and β is the link redundancy parameter.

Denote b_e a binary variable to imply that links e is selected to transmit traffic, here we consider the selected link e as active link. $b_e = 1$ means links e is an active link, while $b_e = 0$ is an unactive link.

2.2 Traffic Model

Denoted the traffic from source switch s destination switch d, and the traffic flow $f_{s,d}$ from s to d passing on link e is defined as $f_{s,d}^e$. Assume u, v are the two endpoints the link e, where $u, v \in V$. In this paper we adopt multipath routing, which the traffic flow is split into multiple sub-flows. Then the flow conservation of all the sub-flows is expressed as follow

$$\sum_d f_{s,d}^{u,v} - \sum_d f_{d,s}^{u,v} = \begin{cases} f^{u,v}, \text{ if } u = s; \\ -f^{u,v}, \text{ if } v = d; \quad u, v, s, d \in \mathcal{V}. \\ 0, \text{ otherwise}; \end{cases} \tag{3}$$

where $N(u)$ is denoted as the set of switches that connected to u.

Assume that the traffic demand between switch s switch d is denoted as $D^{s,d}$. Thus, the traffic flow $f_{s,d}$ should no more than $D^{s,d}$, i.e.,

$$f_{s,d} \leq D^{s,d}, \forall s, d \in \mathcal{V}. \tag{4}$$

2.3 Problem Formulation

In this paper we investigate the routing schedule problem in DCNs by jointly considering the DCNs network throughput and energy consumption. The routing schedule problem is formulated as a mixed integer nonlinear program (MINLP) problem, which has two objectives. The first one is to minimize the network energy consumption, which is expressed as $\min W = \sum_{e \in E} g(x_e)$. The second one is to maximize the network throughput, i.e., $\max T = \sum_s \sum_d f_{sd}, \forall s, d \in \mathcal{V}$. The MINLP problem is formulated as follows.

$$P0 : \min W = \sum_{e \in E} g(x_e), \tag{5}$$

$$\max T = \sum_s \sum_d f_{sd}, \forall s, d \in \mathcal{V}, \tag{6}$$

$$s.t.\ \ x_e = \sum\nolimits_s \sum\nolimits_d f_{s,d}^e, \forall s, d \in V, s \neq d, \tag{7}$$

$$g(x_e) = \sigma + \mu x_e^{\alpha}, 0 \leq x_e \leq \beta C_e, \tag{8}$$

$$\sum_d f_{s,d}^{u,v} - \sum_d f_{d,s}^{u,v} = \begin{cases} f^{u,v}, \textit{if } u = s; \\ -f^{u,v}, \textit{if } v = d; \ u, v, s, d \in \mathcal{V}, \\ 0, \textit{otherwise}; \end{cases} \tag{9}$$

$$f_{s,d} \leq D^{s,d}\ \forall s, d \in \mathcal{V}. \tag{10}$$

In problem ***P0***, the two objectives are maximizing network throughput and minimizing network energy consumption. Constraint (14) states the sum of the traffic demands flows on the link *e*, from the source switch *s* to the destination switch *d*. Constraint (15) presents the energy consumption of each link. Constraint (16) ensures the traffic flow conservation among all the routs. Constraint (16) traffic demand between each switch pair no more than the traffic demands.

2.4 Problem Solution

As analyzed above, the MINLP problem ***P0*** has two conflicting objectives. The first one is minimizing energy consumption, and the second one is maximizing the network throughput. However, there exists a tradeoff between minimizing energy consumption and maximizing the network throughput, i.e., the network throughput enlarges when the energy consumption increases. Thus, it is a key issue to solve the two conflicting objectives problem. Pareto optimal solution [10] is considered as an effective method to solve the two conflicting objectives problem. Here we try to obtain the Pareto optimal solution by transforming problem ***P0*** into problem ***P1***, which has a single objective.

In ***P0***, we first moving the objective (5) into a new constraint,

$$\sum\nolimits_{e \in E} g(x_e) = W_{th}. \tag{11}$$

Changing the value of W_{th} from 0 to W_{max}, we can obtain the maximum of the network throughput when the value of energy threshold W_{th} various. Therefor, problem ***P0*** is transformed into a problem ***P1*** with the objective of maximizing the network throughput.

$$P1 : \max T = \sum\nolimits_s \sum\nolimits_d f_{sd}, \forall s, d \in \mathcal{V}, \tag{12}$$

$$s.t.\ \ \sum\nolimits_{e \in E} g(x_e) = W_{th}, \tag{13}$$

$$x_e = \sum\nolimits_s \sum\nolimits_d f_{s,d}^e, \forall s, d \in V, s \neq d, \tag{14}$$

$$g(x_e) = \sigma + \mu x_e^{\alpha}, 0 \leq x_e \leq \beta C_e, \tag{15}$$

$$\sum_d f_{s,d}^{u,v} - \sum_d f_{d,s}^{u,v} = \begin{cases} f^{u,v}, \text{ if } u = s; \\ -f^{u,v}, \text{ if } v = d; \; u, v, s, d \in \mathcal{V}, \\ 0, \text{ otherwise}; \end{cases} \tag{16}$$

$$f_{s,d} \leq D^{s,d}, \forall s, d \in \mathcal{V}. \tag{17}$$

We try to find the weak Pareto optimal solution of problem ***P0***, i.e., it is no alternative solutions to increase network throughput and reduce energy consumption simultaneously. First, we solve the optimal solution of ***P1***, and then change the energy consumption threshold W_{th} from 0 to the maximum W_{max}, the related T can be derived. Note that, at the beginning the energy consumption is set as 0, which means all the links are un-active in the network, at this scenario, there is no network throughput i.e., $T = 0$. We draw the throughput-network cost curve, as shown in Fig. 1, the network throughput changes when energy consumption increases, especially, when the energy consumption threshold W_{th} exceeds W_s, the network throughout will not modify and keep as T_s. Owing to the Pareto optimal theory, the saturation point (W_s, T_s) means the minimum energy consumption and the maximum network throughput, which is the weak Pareto-optimal point, that is the optimal solution of problem ***P0***.

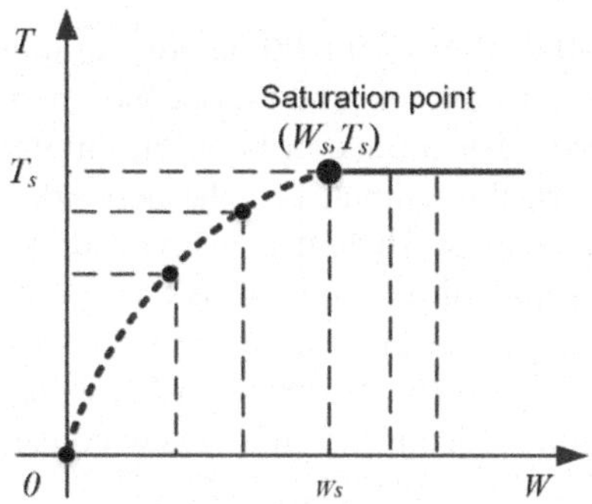

Fig. 1. A diagram of throughput-energy consumption curve.

3 DRL Based Routing Scheme (PEARL)

The problem ***P0*** is MMINLP problem which can be solved by commercial solvers (e.g., CPLEX) based on the Pareto optimal theory, it will take long time and much computing resources, therefore it can not satisfy the dynamic of data center networks, especially for the large scale networks. Here, we design dee**p** r**e**inforcement le**ar**ning based two objectives optimization **r**outing schedu**l**ing (PEARL) scheme based on DRL, due to the advantages of high efficiency and low cost. In this paper PEARL aims to search the optimal action for each state through iterative learning of agent [6].

The structure of PEARL, as described in Fig. 2, contains five parts. In the agent, CNN is used to extract complex mappings for each network states s_t, and then derives the action value, which is denoted as $Q^{\pi_t}(s_t, a_t)$, accordingly. $Q^{\pi_t}(s_t, a_t)$ is defined as a selected routing for DCNs network status s_t. In this paper, to improve network states s_t, we first introduce 2 convolutional neural networks with 2 convolution kernels. The two fully connected layers contains 245 and 100 hidden neurons. The main idea of PEARL is presented as follows.

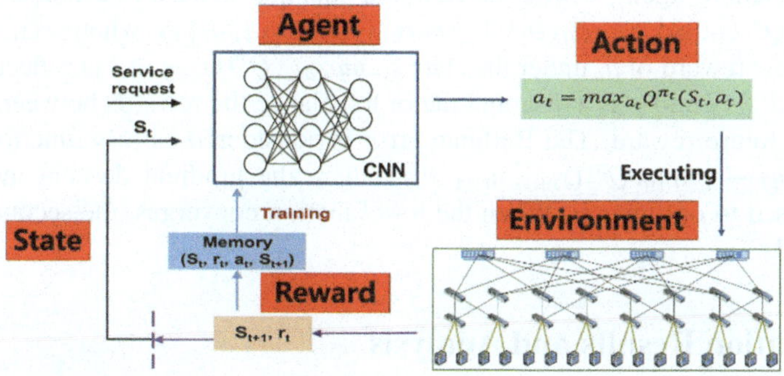

Fig. 2. The proposed DRL-based PEARL model for DCNs.

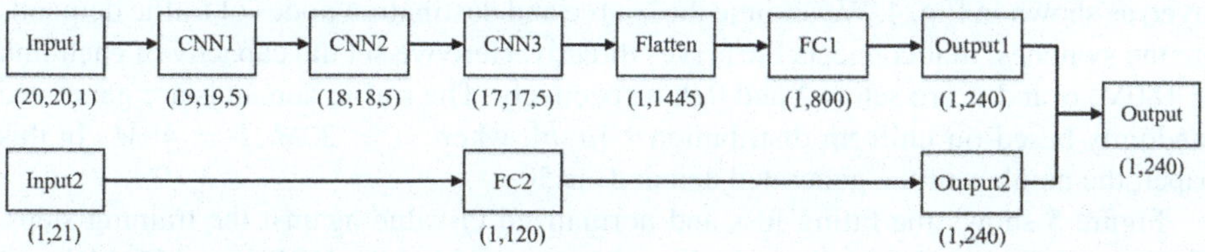

Fig. 3. .

The structure of PEARL contains network environment, agent, network states, routing actions and reward function, which are described as follows. 1) **Environment:** The environment is the data center network, which contains the switches and the links connect them. 2) **Agent:** The agent consists of two convolutional neural networks (CNN) and fully connected neural networks (FC) respectively, where the 2 × 2 convolutional kernels are used in each CNN layer, as shown in Fig. 3. 3) State: The state s_t consists of routing requests (i.e., traffic demands) and links utilization of the data center networks. 4) **Action:** The action a_t describes the selected path for each traffic demand and the action space is 51. 5) **Reward:** The reward function r_t is used to evaluate the performed outcome of action a_t under the state s_t. The agent develops a generalized policy and makes intelligent decisions by repeatedly interacting with the network environment over a number of iterations. The execution steps of DRL based scheme PEARL contains two stages.

The first stage is to generate large amounts of training data. The network state s_t comprises of the traffic demands, the energy consumption of links, and the capacity of links is taken into agent, then the agent returns $Q^{\pi_t}(s_t, a_t)$ of each action a_t according to the policy π_t. Subsequently, the strategy of ε-greedy is used to select action, where a random action a_t is selected with a certain probabilityε, otherwise, the action with the largest value $Q^{\pi_t}(s_t, a_t)$ is selected. Then the selected action a_t^* interacts with the network environment and observes the new network state s_{t+1} as well as reward function r_t. Finally, the data tuple $\{s_t, r_t, a_t, s_{t+1}\}$ is viewed as a new training data and saved in the memory for the next stage.

The second stage is to use the training data to train the agent. Here, PEARL updates total parameters of agent, whose the Bellman equation is used to evaluate each action value, i.e., $Q^{\pi_t}(s_t, a_t) = r(s_t, a_t) + \gamma max_{a_{t+1}} Q^{\pi_t}(s_{t+1}, a_{t+1})$, where $r(s_t, a_t)$ reflects the immediate reward of a_t under the state s_t, $max_{a_{t+1}} Q^{\pi_t}(s_{t+1}, a_{t+1})$ reflects the future value of state s_{t+1} and γ is a discount factor to balance the rewards between immediate reward and future reward. The Bellman error is considered as loss function, i.e., $\varepsilon = Q^{\pi_t}(s_t, a_t, \theta_t) - \gamma \max Q^{\pi_t}(s_{t+1}, a_{t+1}, \theta_t)$, where the gradient descent method (e.g., Adam) is used to optimize it. When the loss function converges, the second stage will come to end.

4 Simulation Results and Analysis

We conduct extensive simulations in a Fat-tree topology, which contains 20 switches, 16 hosts, and 48 links. Since each serve connects only one switch in the aggregation layer, as shown in Fig. 4. We assume the source and destination nodes of traffic demands are the switches, that connects the serves directly. Here we set the capacity of each link as 180M, α and β are set as 2 and 0.9, respectively. The traffic demands are generated randomly based on uniform distribution $U[a, b]$, where $a = 20M$, $b = 40M$. In this paper, the number of the generated demands is 50.

Figure 5 shows the fitting loss and normalized Q-value against the training steps. The loss converges close to 0 gradually, which means that the DRL-based PEARL can be converged after training steps. Figure 6 presents the results of PEARL compared to the routing scheme Floyd-Warshall [9]. Solving problem ***P0*** based on Pareto optimal theory [10], we obtain the relationship between energy consumption and the network throughput. When energy consumption reaches to the saturation point, the network throughput does not increase anymore, based on the Pareto optimality the saturation point is the optimal solution to problem ***P0***. Also, Fig. 6 suggests that the results of PEARL is close to the optimal solution of MINLP, while the performance of Floyd-Warshall is less than PEARL. That is because the heuristic routing scheme is based on shortest distances among the nodes, while PEARL joint considering the energy consumption and the network throughput. Figure 7 shows the execution time versus the number of traffic demands. It shows that the execution time of the PEARL is close to Floyd-Warshall, and much less than the result of MINLP. Therefore, in the dynamic network or in large scale networks, PEARL is the most appropriate compared to the other methods.

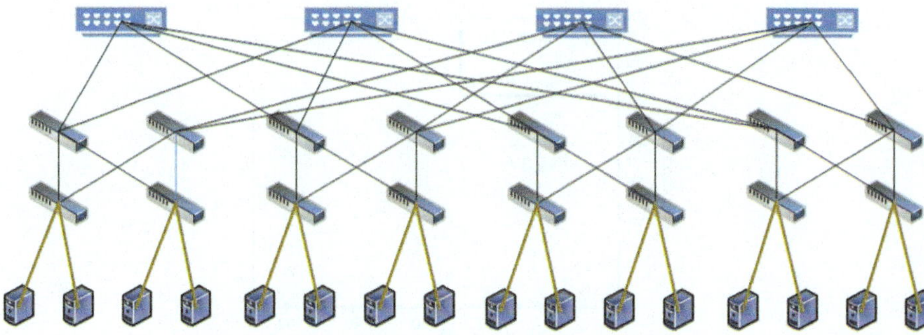

Fig. 4. Simulation network topology of DCNs.

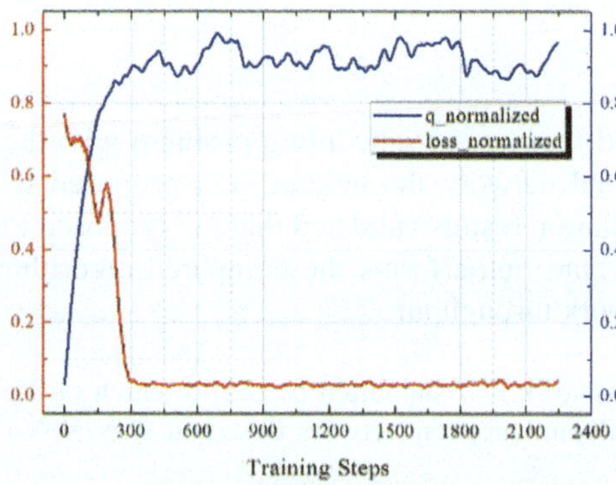

Fig. 5. The convergence property of PEARL.

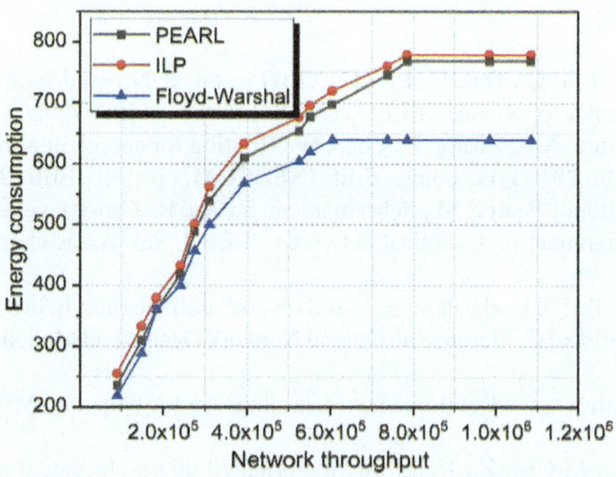

Fig. 6. The network throughput versus energy consumption.

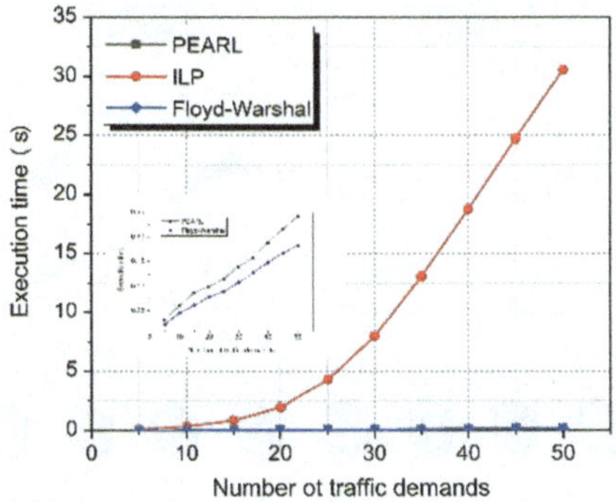

Fig. 7. Execution time.

5 Conclusion

This paper investigated the routing scheduling problem with the joint optimization of energy consumption and network throughput, and proposed the DRL based routing scheme PEARL. Simulation results validated that the proposed PEARL accommodated more traffic demands, and outperforms the compared algorithms in terms of energy consumption and network throughput.

Acknowledgement. This work was supported by the Research on Key Technologies of New Generation Power Data Communication Network Based on SDN/NFV (No. 5700-201952237A-0-0-00).

References

1. Jia, Z., Sun, Y., Liu, Q., Dai, S., Liu, C.: cRretor: an SDN-based routing scheme for data centers with regular topologies. IEEE Access **8**, 116866–116880 (2020)
2. Andrews, M., Anta, A.F., Zhang, L., Zhao, W.: Routing for energy minimization in the speed scaling model. In: 2010 Proceedings IEEE INFOCOM, pp. 1–9. IEEE (2010)
3. Pointurier, Y., Brandt-Pearce, M., Subramaniam, S., Xu, B.: Cross-layer adaptive routing and wavelength assignment in all-optical networks. IEEE J. Sel. Areas Commun. **26**(6), 32–44 (2008)
4. Wang, F., et al.: Traffic load balancing based on probabilistic routing in data center networks. In: 2020 International Conference on Optical Network Design and Modeling (ONDM), pp. 1–3. IEEE (2020)
5. Sutton, R.S., Barto, A.G.: Reinforcement Learning: An Introduction. MIT Press, Cambridge (2018)
6. Xiao, Y., Zhang, J., Gao, Z., Ji, Y.: Service-oriented du-cu placement using reinforcement learning in 5g/b5g converged wireless-optical networks. In: Optical Fiber Communication Conference, p. T4D-5. Optical Society of America (2020)
7. Troia, S., et al.: Machine-learning-assisted routing in ´SDN-based optical networks. In: 2018 European Conference on Optical Communication (ECOC), pp. 1–3. IEEE (2018)

8. Gao, Z., Zhang, J., Yan, S., Xiao, Y., Simeonidou, D., Ji, Y.: Deep reinforcement learning for BBU placement and routing in C-RAN. In: 2019 Optical Fiber Communications Conference and Exhibition (OFC), pp. 1–3. IEEE (2019)
9. Hougardy, S.: The Floyd–Warshall algorithm on graphs with negative cycles. Inf. Process. Lett. **110**(8–9), 279–281 (2010)
10. Gu, Z., Zhang, J., Ji, Y., Bai, L., Sun, X.: Network topology reconfiguration for FSO-based Fronthaul/Backhaul in 5G+ wireless networks. IEEE Access **6**, 69426–69437 (2018)

A Novel Visible Light Communication System Based on a SiPM Receiver

Zhenzhou Deng[1](✉), Liang Ling[1], Yushan Deng[1], Chunlei Han[1,2], Lisu Yu[1], Guojun Cao[3], and Yuhao Wang[1]

[1] School of Information Engineering, Nanchang University, Nanchang 330000, China
zzdeng@ncu.edu.cn
[2] Turku PET Centre, Turku University Hospital and University of Turku, 999018 Turku, Finland
[3] China Institute of Communications, Beijing 100804, China

Abstract. Silicon photomultiplier (SiPM) has several attracting features, which can be helpful in the communication field, such as high photon detection efficiency, fast transient response, excellent timing resolution, and wide spectral range. In this paper, we compare SiPM with photodiode (PD), avalanche photodiodes (APD), and photomultiplier tube (PMT) in terms of the transformation related performance. Based on SiPM and visible light light-emitting diode (LED), we implement visible light communication (VLC) system, and conduct numerical simulation experiment research to convert data transmission problems into signal processing problems. In addition, we also propose the SR algorithm, which compares and analyzes with the PMID and FD algorithms used before. The system throughout ability, transmission rate, and data reconstruction are evaluated. The encouraging results suggest that the SiPM receiver has great application potentials, such as optical wireless communication systems and light fidelity, in which a wide bandwidth of the sensor response is important to enhance the transfer rate.

Keywords: Silicon photomultiplier · Visible light communication · Mean square error

1 Introduction

Self-sustainable green-smart houses [1], optical wireless communication (OWC) [2] and light fidelity (Li-Fi) [3] have attracted considerable attention, in which users always demand large data capacity and multi-functional lighting control during daily life. In contrast to traditional Wi-Fi and fiberoptic communications, the visible light communication (VLC) system [4] is a complementary system with advantages of customizable space, license free, electromagnetic immunity, communication safety and so on. The VLC has found its suitable application scenarios in mobile connection [5], indoor positioning [6], vehicle transportation [7], targeted communication [8], underwater resource exploration [9] and hospital/healthcare applications [10–14].

At this stage, some of the conceptual VLC prototypes have already been implemented in practice for commercial applications. Several worldwide companies including Bytelight, Target, Emart, and Royal Philips NV have successfully guided shoppers to goods

M. Ma (Ed.): ICTCE 2020, LNEE 797, pp. 98–111, 2022.
https://doi.org/10.1007/978-981-16-5692-7_11

based on their position in retail stores, in which light-emitting diodes (LEDs) [15] based VLC are employed to communicate with the camera of smartphones. Other than these applications, intelligent medical care system [16] using VLC has been established and working in several hospitals through the cooperation with the industrial technology research institute (ITRI), which stably and precisely realizes the wireless streaming of therapeutic data and the positioning sensing of medical personnel. According to the report from Grand View Research, all the expected revenue of the global VLC market will reach up to USD 101.30 billion by 2024 [17–21].

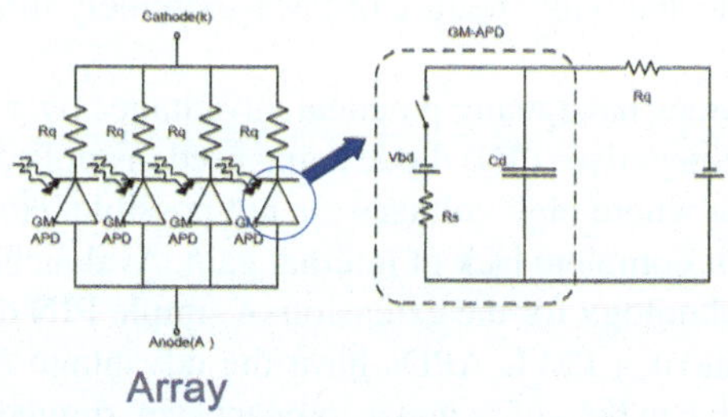

Fig. 1. The inside structure of SiPM. Schematics of a modern SiPM is composed of an array of single photon avalanche diode microcells with passive quenching. It makes SiPM output recovers in short time and has a greater slew rate on the pulse leading edge.

Almost all these VLC applications employ photodiode (PD) to achieve photo-electronic transition and receive data. However, PD still has many shortcomings, such as small area, no internal gain, much lower overall sensitivity (the use of photon counting is limited, usually used for cooling with special electronic circuit photodiodes). Silicon photomultipliers (SiPMs) [22, 23], also identified as multi-pixel photon counters shown in Fig. 1, are a favorable class of semiconductor-based photodetectors addressing the challenge of detecting, timing and quantifying low-light optical signals down to the single-photon counting level. SiPMs offer a highly attractive alternative that extremely increases the detection sensitivity of photo-electronic transition while providing all the benefits of PD [22, 24–27].

In this paper, we focused on a new application of SiPM to the visible light communication field. We discussed the choice cause of SiPM for VLC among different sensors in the second section and developed a SiPM based VLC prototype link in the third section. Then the related performances were investigated in the fourth section. At the end, we provided a summary for the VLC application of SiPM.

2 Comparison with Other Detector Technologies

SiPMs also called SPM, GAPD, MPPC, are a favorable class of semiconductor-based photodetectors addressing the challenge of detecting, timing and quantifying low-light optical signals down to the single photon counting level. SiPMs offer a highly attractive alternative that keeps the low-light detection capabilities of traditional photomultiplier tubes while providing all the benefits of other solid-state devices [28–32]. During the last decade, the SiPM has become more and more popular as photon sensor, especially for applications where the need for a large sensitive area is not an issue but the photon counting capability, pulse bandwidth, and time resolution are important. VLC is only one of the potential applications, in which high pulse bandwidth guarantees the high transfer rate in the light information system.

Up to now, the photomultiplier tube (PMT), a well-established and widely available vacuum tube device, has become the detector of choice for such applications. PMT is generally stable and low noise, but it is bulky and small due to its vacuum tube structure. Magnetic films can also adversely affect them, limiting their suitability for certain applications. The fragile and bulky feature of PMT extremely limited the application of VLC.

Semiconductor devices have many practical advantages over the PMT, and this led to the positive-intrinsic-negative (PIN) diode being used in applications where PMTs are too bulky or delicate, or where high voltages are not possible. However, PIN diodes are severely limited by their complete lack of internal gain. Avalanche photodiodes (APDs) are a relatively new technology for the extension of simple PIN diodes. Whilst the gain may be lower than that of a PMT, APDs have the advantage of a Photon Detection Efficiency (PDE) which can be >65% and a compact size, ruggedness, and insensitivity to magnetic fields. Their main drawbacks are their excess noise and need to trade off performance between noise and time.

The SiPM has high gain and enough PDE (more than 20%), very similar to the PMT, and has the physical benefits of compactness, ruggedness and magnetic insensitivity in common with the PD and APD. In addition, the SiPM achieves its high gain with very low bias voltages (30 V) and the noise is almost entirely at the single photon level. Because the high degree of uniformity between the microcells of the SiPM is capable of discriminating the precise number of photoelectrons detected as distinct, discrete levels at the output node. The ability to measure a well-resolved photoelectron spectrum is a feature of the SiPM which is generally not possible with PMTs owing to the variability in the gain or excess noise.

Table 1. Feature comparison with other detector technologies

Features	PD/PIN	APD	PMT	SiPM
Gain	1	10^2	10^6	10^6
Operational bias	Low	High	High	Low
Readout/electronics	Complex	Complex	Simple	Simple
Form factor	Compact	Compact	Bulky	Compact
Large area available?	No	No	Yes	Yes
Sensitive to magnetic fields?	Yes	Yes	Yes	No

To summarize, among all the photosensors in Table 1, the extremely remarkable performance achieved by SiPM sensors in terms of high photon detection efficiency, fast transient response, excellent timing resolution, and wide spectral range, has made considerable research activities and technological development to be constantly devoted to SiPM devices within the scientific community involved in medical imaging, high-energy physics, and astrophysics. Here, we aim at the realization of specifically designed LED driven VLC architecture to encode, transfer and decode the light or electrical signals containing information and data. Considering the external electronics need to be close to the detector, therefore, SensL SiPM that have an operational bias of 30 V meet the requirements of the extra low voltage directive.

3 SiPM Based VLC Prototype Implementation and Simulation

Our research team has experience in the design and construction of dedicated positron emission tomography (PET) systems using SiPM, starting with the small animal PET called Trans PET [33–36], and then the human PET. In order to improve the performance of the photo detector, a SiPMs array capable of providing high intrinsic spatial resolution and good timing response is designed, and at the same time, a magnetic film can be used. To demonstrate the advance of SiPM detection in VLC, we implement the prototype utilizing SiPM and setup the following simulation experiment.

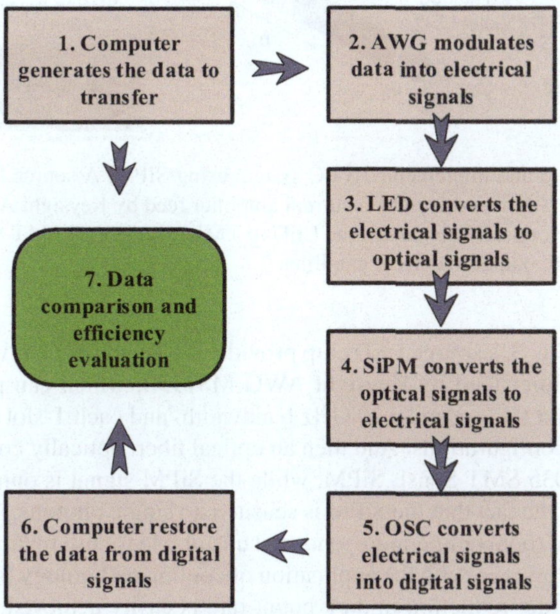

Fig. 2. Operating procedure. In the simulation experiment, there were 7 stages: Computer generates the data to transfer, Arbitrary waveform generator (AWG) modulates data into electrical signals, LED converts the electrical signals to optical signals, SiPM converts the optical signals to electrical signals, Oscilloscope (OSC) converts electrical signals into digital signals, Computer restore the data from digital signals, Data comparison and efficiency evaluation.

3.1 Prototype VLC System Using SiPM

In the simulation experiment, there were 7 stages as showed in Fig. 2. Computer generates the data to transfer, Arbitrary waveform generator (AWG) modulates data into electrical signals. LED converts the electrical signals to optical signals. SiPM converts the optical signals to electrical signals. Oscilloscope (OSC) converts electrical signals into digital signals. Computer restores the data from digital signals, data comparison, and efficiency evaluation.

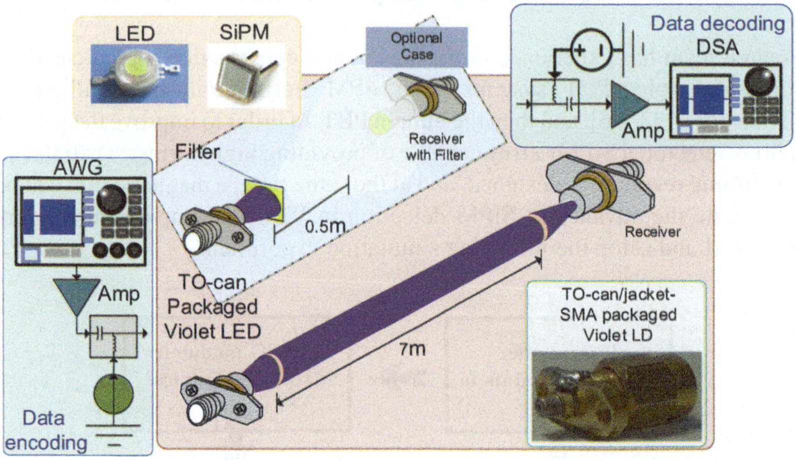

Fig. 3. The proposed and implemented VLC system using SiPM. A source LED chip provided by Nanchang University was driven by a current amplifier feed by Keysight AWG M8195S. And then an optical fiber optically connected the LED to a MicroFB-30035 SMT SensL SiPM, while the SiPM signal were output to current amplifier.

As shown in Fig. 3, a source LED chip provided by Nanchang University was driven by a current amplifier feed by Keysight AWG M8195S, which can provide 65 Giga-Samples per second (GSps) and a 20 GHz bandwidth, and each 1-slot Axie module has 1, 2 or 4 channel configurations. And then an optical fiber, optically connected the LED to a MicroFB-30035 SMT SensL SiPM, while the SiPM signal is output to the current amplifier. Despite the fact that the SiPM is sensitive to single photons, its dark count rate of 100 kHz/mm^2 at room temperature renders it unsuitable for use for applications at very low light levels. However, with the application of cooling technology (such as the SensL MiniSL), a two-order reduction in dark count rate is easily achieved. SensL has more efficient products now, which have more PDE and faster output. We can expect more application when the employing SiPM has better performance. The SiPM is supported with an SMA package connector and its output is directly connected to a Keysight digital storage OSC DSOX6002A with a 50 Ω termination. The OSC is operated with a 6 GHz bandwidth and a 10 GSps sampling rate per channel. At the LED-end, the data are wrapped as electrical pulses serious at different repeat rates: 2 4 5 8 10 20 40 50 80 100 200 400 500 Mega-Samples per second (MSps). The input data from AWG are encoded as different pulse amplitudes. At the SiPM-end, SiPM pulses containing data are digitized by the OSC. Due to the evidential and fast rise of the pulse shape of SiPM, the sent in bits can be easy discriminated after digitized. Since all pulses have tails lasting more than 100 ns, a pulse reconstruction processing is required to obtain the transfer data, which can be verified with the recorded original AWG data.

3.2 Pulse Reconstruction

Caused by the capacitive characteristic, the transferring pulses always pileup with several adjacent ones as shown in Fig. 4. Thus, amplitude of each pulse could be affected when the pileup effect is not considered. To guarantee the received the data consistency and integrality, the pulse reconstruction method is employed. An amplitude modulated SiPM pulse can be modeled as an impulse function $f_i(t) = E_i\delta(t - t_i)$, where $\delta(t - t_i)$ is a shifted Dirac Delta function. For the pulse with index i, its amplitude and arrival time are denoted as E_i and $t_i = i\Delta T$, respectively. Here, $\Delta T = 1/f_r$, where f_r is repeating rate, and ΔT is pulse interval. The overall inputs of the system are a series of photons pulse and can be represented as $f(t) = \sum f_i(t)$. The output signals are waveforms denoted as $p(t)$. The VLC system is then expressed as a linear convolution equation

$$p(t) = f(t) * \varphi(t) + n(t) \tag{1}$$

where $\varphi(t)$ is the system's unit impulse response function, $n(t)$ is the noise and $*$ is the convolution operator. If the pulse interval Δt of adjacent pulses is too close, there will be pileups in $p(t)$. Otherwise, $p(t)$ contains only single events. To retrieve the transferred data E_i, we need to solve the inverse problem of (1) and use the collected signal $p(t)$ to compute $f(t)$.

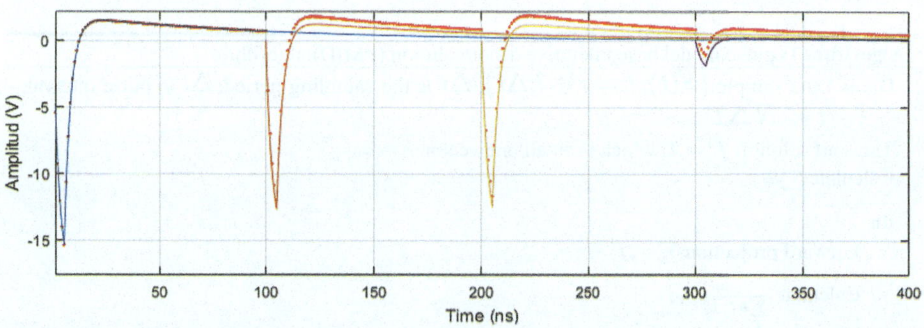

Fig. 4. Pulses pileup is common when the repeat frequency is higher than the reciprocal value of the pulse duration. SiPMs always have capacitive characteristic, then produce pulse tails. The SiPM was support with an SMA package connecter and its output was directly connected to a Keysight digital storage OSC DSOX6002A with a 50 Ω termination.

In practice, signals mentioned above are handled in their discrete forms. We denote $\vec{f} = \{f_j\}, \vec{p} = \{p_l\}$ and $\vec{n} = \{n_l\}$ as the discrete forms of the input, output and noise signal sequences respectively. The system impulse response function is also expressed as a vector $\vec{\varphi} = \{\varphi_r\}$, where $\varphi_r = \varphi(r \cdot \Delta t)$ and Δt is the sampling interval. r, j and l are element indexes in the corresponding vectors. For the convenience of computation, the convolution operation in (1) is rewritten as matrix multiplication. A Toeplitz matrix $H = T\{\vec{\varphi}\}$ is generated from $\vec{\varphi}$ and the convolution (1) is transferred to

$$\vec{p} = H\vec{f} + \vec{n} \tag{2}$$

The element of H is denoted as

$$h_{lj} = \varphi_{l-j} \tag{3}$$

where φ_{l-j} is the $(l-j)$th element in vector $\vec{\varphi}$. Obtaining $\vec{f}$ from (2) is an inverse problem which can be solved by many kinds of well-established methods, including our proposed successive remove (SR) method, pulse model based iterative deconvolution (PMID) method and Fourier deconvolution (FD) method.

In the numerical experiment, the reference method is pulse model based iterative deconvolution [37] and Fourier deconvolution [38]. PMID method employs MLEM iteration to obtain the pulse amplitude. The general form of MLEM algorithm that expresses the updating procedure at the kth iteration is

$$f_j^k = \frac{f_j^{k-1}}{\sum_l h_{lj}} \sum_l h_{lj} \frac{p_l}{\sum_j h_{lj} f_j^{k-1}} \tag{4}$$

where f_j^k denotes jth element of solution $\vec{f}$ at the kth iteration. With enough rounds of iterations, the solution will approach to the original impulse function $\vec{f}$. PMID method can be described as Algorithm 1:

Algorithm 1: pulse model based iterative deconvolution (PMID) algorithm.

Data: Data sample $\{S(t), t = t + k\Delta t\}$, Δt is the sampling period, Δt is pulse interval, $t < N\Delta T$

The start solution $f^0 = \mathbf{1}$, which is an all one vector.

Calculate $\sum_l h_{lj}$.

do

Forward projection $\sum_j h_{lj} f_j^{k-1}$;

Calculate $\frac{p_l}{\sum_j h_{lj} f_j^{k-1}}$;

Backward projection $\sum_l h_{lj} \frac{p_l}{\sum_j h_{lj} f_j^{k-1}}$;

Calculate $\frac{f_j^{k-1}}{\sum_l h_{lj}}$;

Calculate $f_j^k = \frac{f_j^{k-1}}{\sum_l h_{lj}} \sum_l h_{lj} \frac{p_l}{\sum_j h_{lj} f_j^{k-1}}$.

while $r < M$, where M is the loop times.

Output the $E_k, k = 1, 2, 3..., N$. from f^M.

And Fourier deconvolution employs regulated Fourier domain division to calculate Wiener filtering as Eq. 5, which executes an optimal tradeoff between inverse filtering and noise smoothing. It eliminates additive noise while inverting blur. The Wiener filtering is optimal in terms of the mean square error (MSE). FD method can be described as Algorithm 2 as: 2:

$$f_{FD} = \frac{F\{p\}\left(|F\{\varphi\}|^2\right)}{F\{\varphi\}\left(|F\{\varphi\}|^2 + K\right)} \tag{5}$$

where, $F\{\cdot\}$ denotes Fourier transform, and K is a constant, which is defined by the noise property.

Algorithm 2: Fourier deconvolution (FD) algorithm.

Data: Data sample $\{S(t), t = t + k\Delta t\}$, Δt is the sampling period.

Calculate the Fourier Transform of φ;

Calculate the Fourier Transform of p;

Zero padding $F\{\varphi\}$ to the same length as $F\{p\}$;

Calculate $f_{FD} = \frac{F\{p\}(|F\{\varphi\}|^2)}{F\{\varphi\}(|F\{\varphi\}|^2+K)}$;

Output the $E_k, k = 1, 2, 3..., N.$ from f_{FD}.

The VLC system employs an SR method. SR starts with a pulse with ZERO height forward pulses $\{S\{t\}, 0 \leq t < T\}$. Each pulse height estimation follows the remove operation, which reduces the effect of forwards pulses. SR method can be described as Algorithm 3:

Algorithm 3: successive remove (SR) algorithm.

Data: Data sample $\{S(t), t = t + k\Delta t\}$, Δt is the sampling period, ΔT is pulse interval

The 1st SiPM pulse has no pre-component, then $E_1 = \int_0^{\Delta T} S(t)dt$

do

Estimation the pre-component S_k from previous pulses to kth pulse;

Calculate the integral $I(k)$ between $(k-1) \triangle T$ and $k \triangle T$;

Calculate E_k by removing the pre-component $S(k)$, $E_k = I(k) - S(k)$

while $t < N \triangle T$

Output the $E_k, k = 1, 2, 3..., N.$

---

4 Simulation Results

4.1 Single Photon Level Detection

To overcome proportionality lack in APD, the SiPM integrates a dense array of small, electrically and optically isolated Geiger-mode photodiodes. Each photodiode element detects photons identically and independently. The sum of the discharge currents from each of these individual binary detectors combines to form a summed photons output and is thus capable of giving information on the magnitude of an incident photon flux. A spectrum of the same pulse is shown in Fig. 6, and the response to low-level light pulses is shown in Fig. 5.

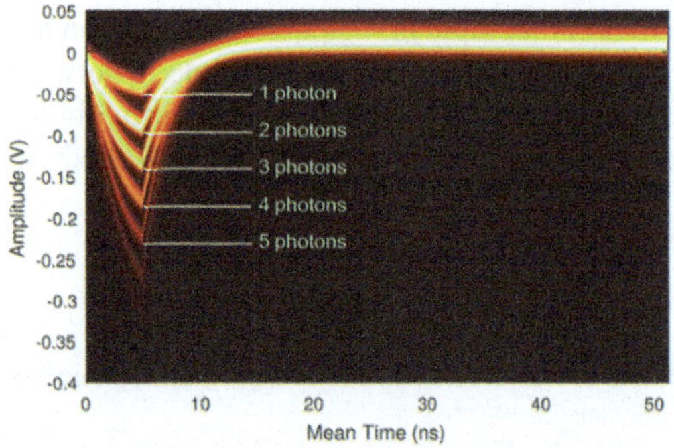

Fig. 5. Responding intensity graph integrated by pulse shapes. Each photodiode element in the array is referred to as a "microcell". Typically, numbering between 100 and 1000 per mm^2, each microcell has its own quenching resistor. The signals of all microcells are then summed to form the output of the SiPM.

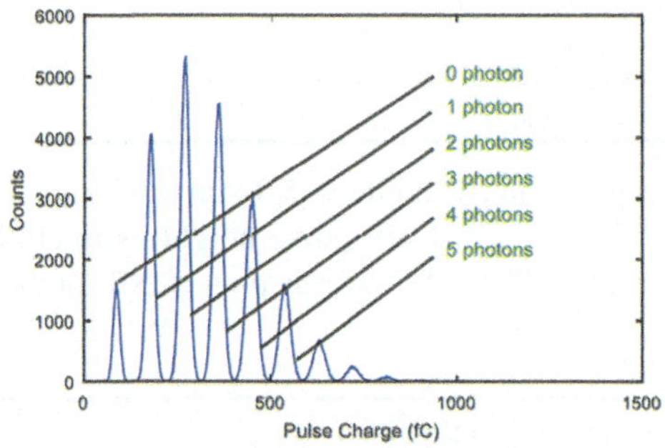

Fig. 6. The single photons' energy spectrum. To overcome proportionality lack in APD, the SiPM integrates a dense array of small, electrically and optically isolated Geiger-mode photodiodes. Each microcell detects photons identically and independently. The sum of the discharge currents from each of these individual binary detectors combines to form a summed photons output, and is thus capable of giving information on the magnitude of an incident photon flux.

4.2 Pulse Shape and SiPM Band Width

In the VLC system, different optical pulses, are fed to SiPM, and generated by LED, and then produced the pulses chain. Transferred data are modulated at the height of optical pulses. We find the allowing data rate without errors depends largely on the single pulse response of the SiPM output. Sharper the SiPM response, more possible transfer rate can be afforded. To know the SiPM response bandwidth, we evaluate the SiPM pulse and its frequency response. Figure 7 shows the impulse response of SiPM, which have a sharp pulse peak with fast leading and recovery edge. And we employ Fast Fourier Transform to obtain its frequency response in Fig. 8. In Fig. 8, we also can see the cut-off frequency that exceeds 100 MHz in Fig. 8, which is significantly better than common photon devices.

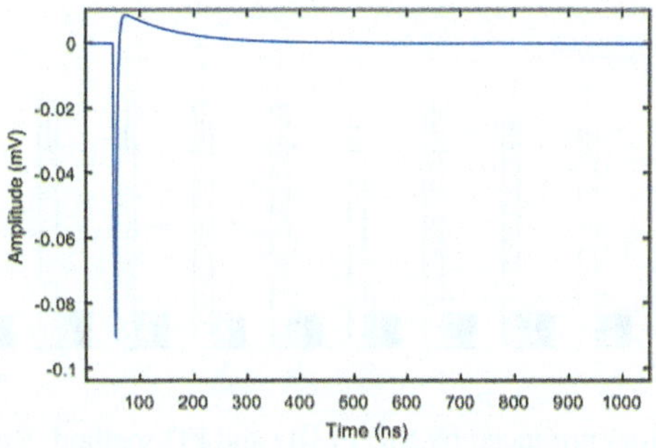

Fig. 7. Mean pulse of the SiPM pulse. Here, the mean pulse is obtained from averaged pulses output by the SiPM on low event rate. Low event rate pulses avoid pileup case. The mean pulse has a sharp pulse peak with fast leading and recovery edge.

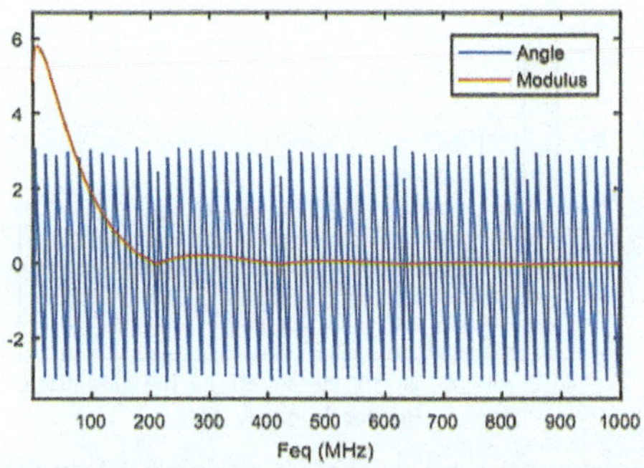

Fig. 8. Frequency property of the SiPM pulse. The VLC system transfer data using the repeating pulses amplitude. So, the frequency width is expected high for more transfer data rate.

4.3 Reconstruction Precision and Required Time

It also can be noted that recovery edge is followed by positive tails in the SiPM pulses. Then pulse aliasing cannot be avoided when the repetition rate is enough to keep the data transfer rate. Thus, we employ three different reconstruction methods to inverse the data input: SR, PMID, and FD. When carrying out the calculation, 100 iterations PMID elapsed more than times which takes more computing time than SR method. And all the 1000 MByte pulse chain with 10 Gsps sample rates is too high to encode by PMID method. To obtain the reconstruction precision of the three methods, we calculate the residual error for just 100 MByte pulse chains.

In Fig. 9, Residual Error of each method is evaluated. Residual Error here means the difference between the received sample and theoretical value calculated from reconstruct data, which is calculated by Eq. 6.

$$r = \left|\vec{p} - H\overrightarrow{f^*}\right|^2 \tag{6}$$

where $f*$ is the vector containing reconstruct data. With the increase of Repeat Rate, Residual Error of three algorithms keeps rising. By contrast, SR and PMID algorithm

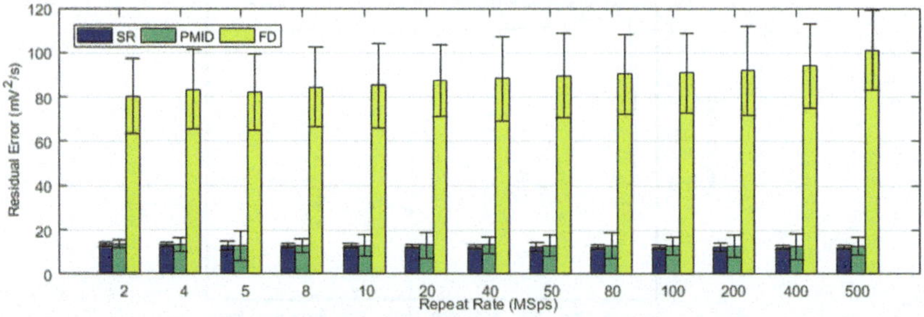

Fig. 9. The Residual Errors produced by SR, PMID and FD method. Residual Error here means the difference between received sample and theoretical value calculated from reconstruct data, which is calculated by Eq. 6.

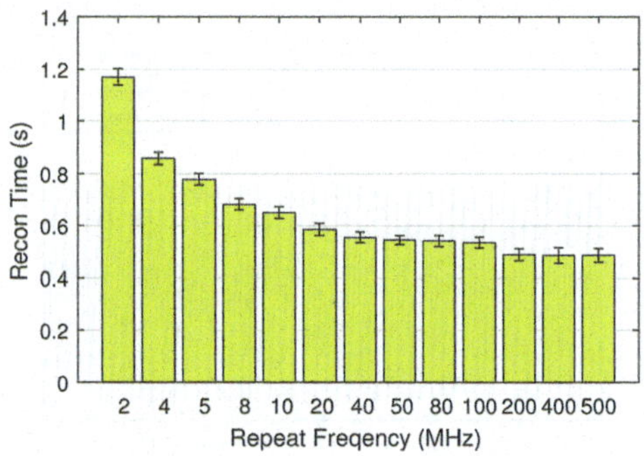

Fig. 10. Required time to restore the pulses chain containing 1 Mbytes using SR method. These time values are calculated on computer workstation. The comparison of elapsed time was based on the hardware of Intel Core i3-3220@3.30 GHz and 8 GB 1600 MHz Kingston DDR3 memory. The program was implemented in MATLAB.

tend to be stable, and is not affected by Repeat Rate too much. We can see that SR and PMID produce one order of magnitude lower residual error than FD. Considering SR processed data free of whole data iteration, we believe the SR is the appropriate method in the practical system. Since each PMID iteration needs 2 convolution operators and 3 vector multiplication operators, it has large computing burden when the iteration number and sample rate increased. We also find the required reconstruction time is affected by the pulse repetition rate. In Fig. 10, we investigate the reconstruction time of SR method. We find the required reconstruction time is decreased as the repetition rate increases. This can be illustrated as that fewer samples need processing when the repetition rate increases.

5 Conclusion

Based on the experience of SiPM application in the imaging system, we constructed a SiPM based VLC system and transferred the data transmission problem to a signal processing problem. Thereafter, we applied the SR algorithm to solve this problem.

Numerical experiments showed that this method provided good signal pulse fitting at a different repetition rate. The pulse reconstruction performance of this method was also better than the previously used PMID and FD methods.

The proposed and implemented system has 4 Gbit/s transfer rates under one 10 GSps DA/AD channels. This system does not require the mean pulse of signals to be in a particular shape, as long as it is fixed. Since different frontend electronics in detection systems may produce different pulse shapes, the adaptability of the proposed method is appropriate in applications. We preliminarily conclude the high transfer rate of SiPM based VLC system attributes the high bandwidth of SiPM signal pulse and single photon counting ability.

In the future work, we will apply the SiPM to different color visible light and multiple voltage threshold digitizers, instead of a single-color LED and the high-speed ADC. Since both Multi-Voltage threshold method and the SR method require the prior knowledge of pulse shape, they could be well-matched. Other iterative algorithms, such as maximum a posterior (MAP) or ordered subset expectation maximization (OSEM), can also be considered in the pulse recovery process.

References

1. Balta-Ozkan, N., Davidson, R., Bicket, M., Whitmarsh, L.: The development of smart homes market in the UK. Energy **60**, 361–372 (2013)
2. Kang, C.H., et al.: High-speed colour-converting photodetector with all-inorganic CsPbBr 3 perovskite nanocrystals for ultraviolet light communication. Light Sci. Appl. **8**(1), 1–12 (2019)
3. Tsonev, D., Videv, S., Haas, H.: Light fidelity (Li-Fi): towards all-optical networking. In: Broadband Access Communication Technologies VIII, vol. 9007, p. 900702. International Society for Optics and Photonics (2014)
4. Rehman, S.U., Ullah, S., Chong, P.H.J., Yongchareon, S., Komosny, D.: Visible light communication: a system perspective—overview and challenges. Sensors **19**(5), 1153 (2019)
5. Ling, R.: The Mobile Connection: The Cell Phone's Impact on Society. Elsevier, San Francisco (2004)
6. Liu, H., Darabi, H., Banerjee, P., Liu, J.: Survey of wireless indoor positioning techniques and systems. IEEE Trans. Syst. Man Cybern. Part C (Appl. Rev.) **37**(6), 1067–1080 (2007)
7. Nizzi, F., et al.: Data dissemination to vehicles using 5G and VLC for Smart Cities. In: 2019 AEIT International Annual Conference (AEIT), Florence, Italy, pp. 1–5. IEEE (2019)
8. Vieira, M.A., Vieira, M., Louro, P., Vieira, P.: Bi-directional communication between infrastructures and vehicles through visible light. In: Fourth International Conference on Applications of Optics and Photonics, Lisbon, Portugal, p. 112070C. International Society for Optics and Photonics (2019)
9. Saeed, N., Celik, A., Al-Naffouri, T.Y., Alouini, M.S.: Underwater optical wireless communications, networking, and localization: a survey. Ad Hoc Netw. **94**, 101935 (2019)
10. Aminian, M., Naji, H.R.: A hospital healthcare monitoring system using wireless sensor networks. J. Health Med. Inform. **4**(02), 121 (2013)
11. Pathak, P.H., Feng, X., Hu, P., Mohapatra, P.: Visible light communication, networking, and sensing: a survey, potential and challenges. IEEE Commun. Surv. Tutor. **17**(4), 2047–2077 (2015)
12. Biton, C., Arnon, S.: Improved multiple access resource allocation in visible light communication systems. Optics Commun. **424**, 98–102 (2018)

13. Islim, M.S., et al.: Towards 10 Gb/s orthogonal frequency division multiplexing-based visible light communication using a GaN violet micro-LED. Photon. Res. **5**(2), A35–A43 (2017)
14. Haas, H.: Visible light communication. In: 2015 Optical Fiber Communications Conference and Exhibition (OFC), Los Angeles, CA, USA, pp. 1–72. IEEE (2015)
15. Schubert, E.F., Gessmann, T., Kim, J.K.: Light emitting diodes. In: Kirk-Othmer Encyclopedia of Chemical Technology (2000)
16. Cahyadi, W.A., Jeong, T.I., Kim, Y.H., Chung, Y.H., Adiono, T.: Patient monitoring using visible light uplink data transmission. In: 2015 International Symposium on Intelligent Signal Processing and Communication Systems (ISPACS), Nusa Dua Bali, Indonesia, pp. 431–434. IEEE (2015)
17. Chi, Y.C., et al.: Violet laser diode enables lighting communication. Sci. Rep. **7**(1), 1–11 (2017)
18. Wang, J., Che, Y., Wang, X., Guo, L., Li, J.: Study of LED layout in indoor visible light communication and performance analysis. In: AOPC 2017: Fiber Optic Sensing and Optical Communications, Beijing, China, p. 1046417. International Society for Optics and Photonics (2017)
19. Khan, L.U.: Visible light communication: applications, architecture, standardization and research challenges. Digit. Commun. Netw. **3**(2), 78–88 (2017)
20. Chi, Y.C., Hsieh, D.H., Tsai, C.T., Chen, H.Y., Kuo, H.C., Lin, G.R.: 450-nm GaN laser diode enables high-speed visible light communication with 9-Gbps QAM-OFDM. Opt. Express **23**(10), 13051–13059 (2015)
21. Janjua, B., et al.: Going beyond 4 Gbps data rate by employing RGB laser diodes for visible light communication. Opt. Express **23**(14), 18746–18753 (2015)
22. Gnecchi, S., et al.: A simulation model for digital silicon photomultipliers. IEEE Trans. Nuclear Sci. **63**(3), 1343–1350 (2016)
23. Corsi, F., et al.: Modelling a silicon photomultiplier (SiPM) as a signal source for optimum front-end design. Nucl. Instrum. Methods Phys. Res., Sect. A **572**(1), 416–418 (2007)
24. Eraerds, P., Legré, M., Rochas, A., Zbinden, H., Gisin, N.: SiPM for fast photon-counting and multiphoton detection. Opt. Express **15**(22), 14539–14549 (2007)
25. Braga, L.H., et al.: A fully digital 8 × 16 SiPM array for PET applications with per-pixel TDCs and real-time energy output. IEEE J. Solid-State Circuits **49**(1), 301–314 (2013)
26. Powolny, F., et al.: Time-based readout of a silicon photomultiplier (SiPM) for time of flight positron emission tomography (TOF-PET). IEEE Trans. Nuclear Sci. **58**(3), 597–604 (2011)
27. Várkonyi, G., McKendry, J.J., McAlinden, N., Dawson, M.D., Mathieson, K.: Data transmission for high-bandwidth neural interfacing using visible light communication. In: 2016 IEEE Photonics Conference (IPC), Waikoloa, HI, USA, pp. 41–42. IEEE (2016)
28. Yepes-Ramirez, H., et al.: Development of a high performance characterization setup for SiPMs and MPGDs towards their integration in mid-large scale systems. In: Proceedings of Science (2016)
29. Castañeda, L.F., Castaño, J.F., Gutierrez, R.M., Hernandez, A.I., Yepes-Ramirez, H.: High-speed visible light communication system based on SiPM. In: Proceedings of the 38th International Conference on High Energy Physics (ICHEP 2016), Chicago, p. 842 (2016)
30. Konstantinou, G., Ali, W., Chil, R., Cossu, G., Ciaramella, E., Vaquero, J.J.: Experimental demonstration of an optical wireless MRI compatible PET/SPECT insert front-end. In: 2016 IEEE Nuclear Science Symposium, Medical Imaging Conference and Room-Temperature Semiconductor Detector Workshop (NSS/MIC/RTSD), Strasbourg, pp. 1–4. IEEE (2016)
31. Buzhan, P., et al.: Large area silicon photomultipliers: performance and applications. Nucl. Instrum. Methods Phys. Res., Sect. A **567**(1), 78–82 (2006)
32. Bencardino, R., Eberhardt, J.E.: Development of a fast-neutron detector with silicon photomultiplier readout. IEEE Trans. Nucl. Sci. **56**(3), 1129–1134 (2009)

33. Deng, Z., Xie, Q.: Empirical Bayesian energy estimation for multi-voltage threshold digitizer in PET. In: 2013 IEEE Nuclear Science Symposium and Medical Imaging Conference (NSS/MIC 2013), Seoul, Korea, pp. 1–5. IEEE (2013)
34. Xie, Q., et al.: Potentials of digitally sampling scintillation pulses in timing determination in PET. IEEE Trans. Nuclear Sci. **56**(5), 2607–2613 (2009)
35. Deng, Z., Xie, Q.: Quadratic programming time pickoff method for multivoltage threshold digitizer in PET. IEEE Trans. Nucl. Sci. **62**(3), 805–813 (2015)
36. Deng, Z., Liu, Y., Wu, Z., Xie, Q.: Maximum likelihood solution of arrival time for the event pulse modeled by filtered inhomogeneous poisson process in PET detectors. J. Nucl. Med. **56**(Suppl. 3), 1842 (2015)
37. Deng, Z., Xie, Q., Duan, Z., Xiao, P.: Scintillation event energy measurement via a pulse model based iterative deconvolution method. Phys. Med. Biol. **58**(21), 7815 (2013)
38. Nagaoka, R., Yoshizawa, S., Umemura, S.I., Saijo, Y.: Basic study of improvement of axial resolution and suppression of time side lobe by phase-corrected Wiener filtering in photoacoustic tomography. Jpn. J. Appl. Phys. **57**(7S1), 07LD11 (2018)

Research on Peak-Detection Algorithms of Fiber Bragg Grating Demodulation

Xiangyu Guo(✉)

School of Precision Instrument and Opto-Electronics Engineering,
Tianjin University, Tianjin 300072, China

Abstract. Peak-detection algorithm is a crucial part of fiber grating signal demodulation. It plays a decisive role in the accuracy, speed and anti-noise performance of the entire system. This article introduces the research focusing on peak-detection algorithms and summarizes the common peak-detection algorithms in recent years into three types: traditional direct peak-detection algorithms, classic fitting peak-detection algorithms, and innovative improved algorithms. It mainly includes direct peak-detection algorithm, centroid detection algorithm, polynomial fitting method, gauss fitting algorithm, cubic spline interpolation algorithm, adaptive semi-peak-seek algorithm and Steger peak detection algorithm. The principles of various peak-finding algorithms are explained and the characteristics and performance of each algorithm are comprehensively compared. It is convenient to choose the optimal demodulation algorithm in the future research tasks according to different actual conditions and requirements.

Keywords: Fiber Bragg grating · Peak-detection algorithm · Fiber Bragg Grating Sensing

1 Introduction

Fiber grating is a kind of optical device with grating structure. Its main working principle is to make the external photons received by the optical components and the germanium ions inside the fiber core interact, so that the refractive index is permanently changed, and a spatial phase grating is formed inside. Among them, the fiber Bragg grating (FBG) sensor is a new type of fiber wavelength modulation sensor with a high frequency of use and wide application popularity in the future. The principle is to judge by sensing the change of external physical quantity and the shift amount of the center wavelength of the reflection spectrum collected by the spectrum demodulator and completing the analysis and processing through this judgment [2]. Due to its anti-electromagnetic interference, small size, easy reuse and high sensitivity, it can directly or indirectly detect physical quantities such as strain, temperature, displacement and vibration, making it suitable for optical fiber sensing, civil engineering, aerospace, and composite materials, medical equipment and other fields have a wide range of applications [1]. Due to the adverse effects of nonlinear temperature drift, noise interference, spectral overlap and distortion

M. Ma (Ed.): ICTCE 2020, LNEE 797, pp. 112–122, 2022.
https://doi.org/10.1007/978-981-16-5692-7_12

of optical devices and the shortcomings of traditional fiber grating demodulation technology, such as imprecision, speed and instability, high-precision demodulation of its wavelength signal have become a very important problem. To summarize and compare the factors that affect the accuracy of FBG wavelength demodulation, it can be divided into two technical aspects: hardware and software. The software part mainly uses the peak-detection algorithm to process the FBG spectrum to obtain its center wavelength. The peak-detection algorithm will greatly affect the detection accuracy of the entire sensor system, so this article mainly analyzes traditional direct peak-detection algorithms, classic fitting peak-detection algorithms, and innovative improved algorithms for in-depth discussion.

2 Research Focus of Peak-Detection Algorithm

The peak-detection algorithm is simply an algorithm used to find the peak point. The main function of the peak-detection algorithm is to find the peak points in a large number of discrete discontinuous data collected. In practical applications, the peak-finding algorithm must have the basic requirements of good accuracy and fast speed. Therefore, the impact on the application of the peak-detection algorithm is mainly in the following two aspects:

- Accuracy influence. When the fiber grating is working, the optical device itself, the external working environment and the internal system hardware will introduce a lot of noise to the data collected by the demodulation system, which will cause the optical power to change, and further will cause the center wavelength of the FBG reflection spectrum to deform and jitter. In the end, the accuracy of the subsequent peak-finding is greatly reduced, and a large amount of errors are generated, which greatly affects the applicability of the peak-detection algorithm [2].
- Speed influence. Because of the practicability of fiber Bragg grating demodulation system, in order to be able to be used in engineering projects, the speed of the whole demodulation system must be fast enough, and the calculation and processing speed of peak-detection algorithm is a crucial part of determining the running speed of the whole system. The more iterative operations are introduced in the algorithm, the higher delay of the whole system, which also greatly affects the real-time performance of the demodulation system [2].

3 Wavelength Peak-Detection Algorithm

There are three types of wavelength peak-detection algorithms: traditional direct peak-detection algorithm, classic fitting peak-detection algorithm, and innovative improved algorithm for discussion. The specific classification of the algorithm is shown in Fig. 1. Among them, the traditional direct peak-detection algorithm mainly includes direct peak-detection algorithm and centroid detection algorithm. Classic fitting peak-detection algorithm mainly includes polynomial fitting algorithm, Gauss fitting algorithm and cubic spline interpolation algorithm. The innovative improved algorithm is mainly to improve the first two types of algorithms in different aspects to adapt to certain specific requirements or to strengthen the performance of the algorithm for a certain aspect, and its change forms are relatively diverse. Due to space limitations, this article only introduces the more common adaptive semi-peak-seek algorithm and the steger peak detection algorithm.

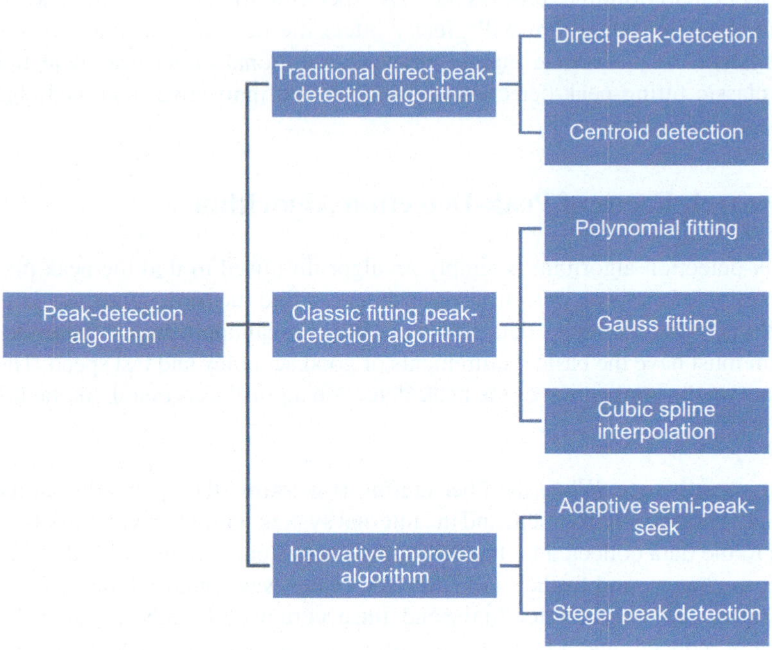

Fig. 1. Classification of peak-detection algorithms

3.1 Traditional Direct Peak-Detection Algorithm

The traditional direct peak-detection algorithm does not change or process the shape of the FBG reflection spectrum, and directly processes the data to obtain the peak range. Compared with other algorithms, the calculation amount of this type of algorithm is relatively small, so its calculation speed is relatively fast, but the system noise has a greater impact on it, and its demodulation accuracy is often relatively low, so it needs to be peak-searching in an ideal environment.

3.1.1 Direct Peak-Detection Algorithm

The direct peak-detection algorithm is also called the linear interpolation and differentiation algorithm, which mainly performs peak splitting on the collected waveform curve data. Peak splitting refers to setting a threshold to separate multiple peak waveform areas, and the original peak position is determined by the position of the zero point of the first-order numerical differential [2]. In specific engineering use, an initial valley value and an initial peak value can be preset for the data, and the comparison method is used to sequentially compare the collected fiber grating reflection signals, record the position less than the preset valley value and temporarily store the data as Bottom value, record the position greater than the preset peak value and temporarily store its value as the peak value. When the following conditions are met at the same time: (1) The peak position is greater than the valley position (2) The difference between the peak and the valley is higher than a preset threshold (3) The peak is higher than the current scan signal by the same threshold. At this time, it can be judged that the peak value is a true peak value, the current peak position is recorded in the peak position array, and the valley value and peak size are reset to the initial value, and the above steps are repeated to read the array until the entire array is read [4].

3.1.2 Centroid Detection Algorithm

The centroid detection algorithm is also called the power weighted average algorithm in the application of FBG sensor signal processing. This algorithm considers from the perspective of quality and uses a quality "centroid" to correspond to the collected curve waveform data. The position vector in the mass point system corresponds to the abscissa of each point in the reflection spectrum, and the mass in the mass point system corresponds to the amplitude of each point, and then the weighting coefficient is assigned to each data to reduce the impact of accidental measurement errors. Finally, the weighted average of all data is used to find the peak position [2]:

$$x_0 = \frac{\sum x_i y_i}{\sum y_i} i = 1, 2, \ldots, N x_0 = \frac{\sum x_i y_i}{\sum y_i} i = 1, 2, \ldots, N \quad (1)$$

Among them, y_i is the ordinate of the mass point, x_i is the wavelength corresponding to each point, x_0 is the center wavelength of the FBG, and N is the number of sampling points.

In actual engineering applications, the reflected energy of FBG is basically concentrated in its reflection passband. The reflected light power can be used as a weighting coefficient to calculate the weighted average of wavelengths to obtain the distribution of the reflected light power in the wavelength direction, thereby obtaining the center wavelength of the FBG power reflection spectrum, and reflecting the power distribution of the entire reflection spectrum [3].

3.2 Classic Fitting Peak-Detection Algorithm

Classic fitting peak-detection algorithms usually need to use functions to process the FBG spectrum. This algorithm can filter out most of the noise to obtain a higher detection accuracy, but it has a larger amount of calculation compared with other algorithms. And the waveform curve of the collected data conforms to the specific form, so its performance will be affected by the spectrum type. The larger the waveform shape and function deformation of the collected points, the lower the peak searching accuracy of the algorithm.

3.2.1 Polynomial Fitting Algorithm

The collected curve waveform data generally appears in a form similar to a certain function curve, so the discrete data collected by function fitting can be used. The principle is to use a polynomial fitting function to fit the spectral data, and the reflection spectrum sampling points are obtained by peak splitting the acquired waveform curve data. Using the least squares method as the judgment method, the fitting polynomial is obtained, and then the extreme points are calculated, so as to obtain the corresponding peak points [5].

Assuming that the fitted polynomial is:

$$p_n(x) = a_0 + a_1 x + \cdots + a_n x^n \tag{2}$$

The first-order differential analytic formula is:

$$p_n'(x) = a_1 + 2a_2 x + \cdots + n a_n x^{n-1} \tag{3}$$

Then the solution of the equation is the peak position of the corresponding fitting function:

$$a_1 + 2a_2 x + \cdots + n a_n x^{n-1} = 0 \tag{4}$$

3.2.2 Gauss Fitting Algorithm

For FBG reflectance spectrum which accords with Gauss curve distribution, the peak position can be obtained directly by Gauss fitting algorithm. The approximate expression of Gaussian function is as follows:

$$I(x) = a\exp\left[-\left(\frac{x - x_0}{\sqrt{2}\sigma}\right)^2\right] \tag{5}$$

Where a is the amplitude of the reflection spectrum, x is the wavelength of the FBG, x_0 is the center wavelength of the FBG, and σ is the 3dB bandwidth of the reflection spectrum. According to the least square method to find the sum of squared deviations S, the best fit is required, which is to make the sum of squared deviations S take the minimum [2]:

$$S = \sum_{i=1}^{n} [I_i - I(x_i)]^2 = \min \tag{6}$$

The peak point is the FBG center wavelength x_0.

3.2.3 Cubic Spline Interpolation Algorithm

Interpolation is one of the commonly used methods in the data processing. The piecewise low-order interpolation has the advantages of relatively less calculation, continuous interpolation curve and approximation to the function, but there will be "sharp points" at the piecewise point of the piecewise difference polynomial, which makes the fitting smoothness worse. Although the high-order interpolation calculation solves the problem of fitting smoothness, its calculation is too complicated, the calculation time required is very long, and the Runge phenomenon may occur. In order to improve the overall smoothness, cubic spline interpolation constructs an interpolation function with a second-order continuous derivative as a whole under the condition that only the function value at the interpolation point is given. The principle is that after cutting the sampled data, performing cubic spline interpolation on it, and calculating the numerical integration, and the point where the difference is less than a preset small value is the highest point [5].

For example, WangZi-shuo [3] approximates the peak data in the form of the following formula:

$$f = p\sum_{i=1}^{n-1} w_i(y_i - f(x_i))^2 + (1 - p)\int_{x_0}^{x_{n-1}} \lambda(x)\left(f^n(x)\right)^2 dx \tag{7}$$

Where p is the balance parameter, when $p = 0$, it is equivalent to linear fitting. When $p = 1$, it is equivalent to cubic spline interpolation. p is chosen in the interval [0, 1] to make the fitted curve smooth and close to the data points. The closer p is to 0, the smoother the fitted curve is. The closer p is to 1, the closer the fitted curve is to the data point.

3.3 Innovative Improved Algorithm

Innovative improved algorithms have become more and more widely used in recent years with the development of optical fiber sensing systems in the direction of precision and high speed. The main reason is that most of these algorithms are to improve the above two types of algorithms to adapt to some special occasions, such as requiring the system to have extremely high anti-noise performance, be able to perform dynamic measurements, and obtain extremely high demodulation accuracy.

In 2016, ChenYong [9] and others proposed an exponential modified Gauss (EMG) fitting peak-detection algorithm, and proved that this algorithm has high demodulation accuracy and can overcome the asymmetry of the FBG reflection spectrum. In 2017, Jiang Hong [10] and others used Karhunen-Loeve transform for FBG demodulation, which proved that this method can achieve high-precision demodulation. In 2020, TongGuowei et al. [7] proposed a FBG demodulation method based on meta-heuristic algorithm, which uses the center wavelength search algorithm of evolutionary algorithm. This algorithm can simultaneously ensure the accuracy of demodulation and realize multi-spectral overlap recognition. In the same year, Jiang Hong [8] and others introduced the weight factor into the least squares fitting algorithm to implement the Weighted Least Squares fitting algorithm (WLS), and optimized the Gaussian curve fitting coefficient.

Then the Asymmetric Gaussian correction (AG) is used to correct the center wavelength of the positioning, and the WLS-AG algorithm is proposed. This algorithm can resist noise interference for high-precision peak finding, especially in the case of high signal-to-noise ratio. Due to limited space, this article will mainly introduce the widely used adaptive semi-peak-seek algorithm and the steger peak detection algorithm. Both of these algorithms consider the impact of asymmetric peak spectrum on demodulation.

3.3.1 Adaptive Semi-Peak-Seek Algorithm

This kind of algorithm is often used for high-speed demodulation, and a sampling threshold is usually defined in the algorithm design to control the system. In this way, some low-power clutter and short-wavelength sidelobes at the bottom of the signal can be cut to filter out external signal interference caused by power supply, wiring, etc. When the sampled value is higher than the threshold, the system works in kurtosis, and when the sampled value is less than the threshold, the system will not analyze it. "Start position" is recorded as the system is in the ascending state and the sampling value is greater than the threshold for the first time, and "end position" is recorded as the system is in the descending state and the current value is less than the threshold for the first time.

The mathematical expectation of the sampled data E(y) and the quantile value $y_{75\%}$ will be able to be calculated. The quantile calculation can select different threshold quantiles according to the actual situation. The half-peak detection threshold Y_{avr} is obtained after averaging the above two values, and the abscissas of the two sampling points corresponding to Y_{avr} are taken as the position values X_m and X_n. Since X_m and X_n are located at the rising and falling edges of the spectrum at this time, the sensitivity of signal detection is high, and the interference at the top and bottom of the curve has little effect on it. The peak position can be considered as the middle position X_{avr} of X_m and X_n [15].

$$E(y) = \sum y_i^* P(y = y_i)$$

$$P(y < y_{75\%}) = 75\% \tag{8}$$

3.3.2 Steger Peak Detection Algorithm

The image processing algorithm has the advantages of high precision in target recognition and position extraction, parallel and fast processing by special chip. Steger peak detection algorithm uses the image processing algorithm to identify the half-waist position of the spectrum in the peak finding process, and obtains the precise position of the spectral peak and the asymmetry correction amount according to the relative relationship between the half-waist position and the position of the spectral peak. It can improve the anti-noise performance and peak-finding accuracy of the demodulation system.

Firstly, the reflection spectrum f(x) of the FBG with the Gaussian template is convolved and its first and second order differential forms:

$$r(x, \sigma) = g_\sigma(x) * f(x)$$

$$r'(x, \sigma) = g'_\sigma(x) * f(x)$$

$$r''(x, \sigma) = g''_\sigma(x) * f(x) \tag{9}$$

Among them, $g_\sigma(x) = \frac{1}{\sqrt{2\pi\sigma}}\exp(-\frac{x^2}{2\sigma^2})$, σ is the bandwidth of the Gaussian function. When $\sigma = N_{3dB}/8$, the half waist position of the spectrum can be located, and N_{3dB} is the number of sampling points in the spectrum bandwidth; usually the size of the Gaussian template is $8[\sigma] + 1$, [] is the rounding symbol.

Then a second-order Taylor expansion can be performed on the x_0 neighborhood of r [11]:

$$r(x_0 + \delta x) = r(x_0) + r'(x_0)\delta x + \frac{1}{2}r''(x_0)\delta x^2 \tag{10}$$

According to the derivative at the peak point being zero, the sub-step correction amount of the peak position is obtained:

$$\delta x = -\frac{r'(x_0)}{r''(x_0)}, \delta x \in [-0.5l_{step}, 0.5l_{step}] \tag{11}$$

The center position can be expressed as:

$$l = \delta x + x_0 \tag{12}$$

3.4 Comprehensive Comparison of Commonly Used Algorithms

The performance and characteristics of common peak finding algorithms are comprehensively compared [1, 4–6, 12–14], and the result is shown in Fig. 1.

Classification	Peak-detection algorithm	Features	Performance
Traditional direct peak-detection	Direct peak-detection	The shape of FBG reflection spectrum is not changed	The amount of calculation is the smallest and the required demodulation time is the shortest, but the demodulation accuracy is very low, and there is almost no anti-noise performance. Among them, the centroid detection algorithm has a more complicated calculation amount and a slightly higher demodulation accuracy

(continued)

(*continued*)

Classification	Peak-detection algorithm	Features	Performance
	Centroid detection		
Classic fitting peak-detection	Polynomial fitting Gauss fitting Cubic spline interpolation	Use function to process FBG spectrum to change its shape	The algorithm uses functions to fit the spectrum, requires the spectrum to conform to a specific shape, and increases the amount of calculation, but the demodulation accuracy and anti-noise performance are improved. Compared with the traditional direct peak-detection algorithm, the overall demodulation accuracy is higher, the noise resistance is better, but the demodulation speed is slower. Among them, the Gauss fitting has the highest demodulation accuracy, the best noise resistance and the longest calculation time. The demodulation accuracy of cubic spline interpolation fitting method is similar to that of Gauss fitting, and the demodulation speed is general. It has more advantages than Gauss fitting in dynamic fitting. The accuracy of polynomial fitting demodulation is low, and the demodulation speed is faster
Innovative improved algorithm	Adaptive semi-peak-seek	Identify spectra by setting a threshold	Compared with Gauss fitting, the adaptive semi-peak-seek has better accuracy, speed and noise immunity

(*continued*)

(*continued*)

Classification	Peak-detection algorithm	Features	Performance
	Steger peak detection	Use image processing algorithms to identify spectra	The Steger peak has the highest demodulation accuracy and the best anti-noise performance among the listed algorithms. Its demodulation speed is slightly slower than the adaptive semi-peak-seek, but it is significantly better than the Gauss fitting

4 Conclusion

In recent years, the application of fiber grating sensing technology has become more and more widespread. Among them, the peak-detection algorithm plays an important role in the demodulation of fiber grating signals. The improvement and innovation of the peak-detection algorithm has become the current research focus of many scholars. This article first introduces the research focus of peak-detection algorithms in recent years, then summarizes the common peak-detection algorithms into three categories, and finally analyzes and compares the advantages and disadvantages of the above-mentioned algorithms. In future research, the peak-detection algorithm will continue to develop in the direction of higher accuracy, stronger anti-noise, and better real-time performance. Artificial intelligence algorithms may become a new research hotspot. The optimization of artificial intelligence algorithms and the application problems in small embedded systems are urgently needed for further research.

Acknowledgement. I appreciate my college which gives me a comfortable learning atmosphere. Second, I would like to show my deepest gratitude to my supervisor, Miss Sun, who has walked me through all the stages of the writing of this thesis. Without her illuminating instruction and patience, this thesis could not have reached its present form. I am also greatly indebted to all my teachers who have helped me to develop fundamental and essential academic competence.

References

1. Yao, G., Yin, Y., Li, Y., Fan, H.: Summary of research on high precision fiber grating wavelength demodulation method. Study Opt. Commun. 1–14 (2021)
2. Guo, X.: The Design and Research of the Fiber Grating Demodulating System and Algorithm. North University of China (2017)

3. Wang, Z.: Research on the Demodulation Method of Optical Fiber Grating Sensing Signal. North China Electric Power University (2011)
4. Sun, J., Huang, J., Gu, H., Li, S., Yang, G.: Direct peak-detection algorithm and centroid detection algorithm in Tunable F-P filter based demodulation system. J. Sichuan Ordnance **32**(06), 83–85 (2011)
5. Liu, Q., Wang, H.: Research on the peak-detection algorithm in the high-frequency demodulation for the fiber Bragg grating. J. Wuhan Univ. Technol. **32**(06), 59–61+88 (2010)
6. Yang, Z., Ju, W., Li, Y.: Research advances in high precision demodulation of FBG wavelength. Study Opt. Commun. 1–9 (2020)
7. Tong, G., Liu, B., Huang, L., Chen, C., Xu, H.: FBG demodulation method based on meta-heuristic algorithm. Study Opt. Commun. 1–8 (2021)
8. Jiang, H., Guo, Y., Zheng, X., Liu, P., Zhou, S.: Research on an improved spectral peak seeking algorithm with high precision. Study Opt. Commun. (01), 33–37 (2020)
9. Chen, Y., Yang, X., Liu, H., Yang, K., Zhang, Y.: Exponential modified Gaussian fitting peak-detection algorithm for FBG sensing signal processing. Spectrosc. Spectral Anal. **36**(05), 1526–1531 (2016)
10. Jiang, H., Li, J., Zhang, S., Wang, X., Wang, F.: Research on fiber grating demodulation algorithm based on K-L transform. Laser Infrared **47**(08), 1029–1032 (2017)
11. Wang, Q., Yang, Y.: A FBG spectrum peak detection technique based on Steger image algorithm. Acta Optica Sinica **34**(08), 133–138 (2014)
12. Guo, X., Yang, Q.: Research on an asymmetric Gaussian fitting peak-detection algorithm based on FGA state machine. Electron. Des. Eng. **26**(01), 161–165 (2018)
13. Zhang, J., Xiong, Y., Wu, Z., Li, W.: Wavelength demodulating algorithm of FBG dynamic tuned by DFB laser. J. Harbin Univ. Sci. Technol. **24**(02), 139–143 (2019)
14. Li, N., Wang, D., Wang, Y., Bai, Q., Zhou, H., Jin, B.: Research on influence of curve fitting algorithm on demodulation performance of fiber Bragg grating sensor. Chin. J. Sens. Actuators **32**(05), 711–714 (2019)
15. Liu, Q., Cai, L., Li, Z., Tang, Z., Du, F., Zhao, M.: Research on peak-detection algorithm for high-speed and high-precision FBG demodulation. J. Optoelectron. Laser **23**(07), 1233–1239 (2012)

Maintenance of Utilities Communication Equipment. From Normal to Excellence

Seena Zarie(✉)

Dubai Electricity and Water Authority (DEWA), Dubai, UAE
seena@dewa.gov.ae

Abstract. Excellence is a current focus for developed countries. Utilities, which are one of the critical sectors, need to apply excellence in order to deliver the required results to all their stakeholders. Communication network is a main element for any utility. Maintaining communication network requires additional efforts nowadays, which calls the need of shifting from normal to excellent maintenance. Applying excellent maintenance will ensure the availability and sustainability of the communication network, which will ensure utilities' continuous operations. This paper aims to highlight the ways to achieve excellence maintenance for communication equipment in the utility sector. The study is based on several challenges and studies done by DEWA (Dubai Electricity and Water Authority) communication engineers during maintenance stage. The study concludes that it is essential to take care about all stages in order to apply excellent maintenance.

Keywords: Communication · Utility · Excellence · Maintenance

1 Introduction

Excellence is defined as being best of the best superior to your competitors [1]. It is important to have the excellence culture in organizations among employees. One of the major points in this excellence culture is that all factors in all stages – whether it is major or minor - are important.

Maintenance is a critical part of a facility's operation. Properly maintained equipment and processes are necessary to keep the facility functioning at its optimum capability [2]. Maintenance is an essential process for any organization in order to keep its operations.

Communication Network, which is defined as transmission, reception and processing of information between two or more locations using electronic circuits [3] is one of the critical networks for utilities that needs continuous maintenance. This network consists of interconnection of a number of nodes made up of intelligent processors. The primary purpose of these nodes is to route data through the network [4].

M. Ma (Ed.): ICTCE 2020, LNEE 797, pp. 123–134, 2022.
https://doi.org/10.1007/978-981-16-5692-7_13

This communication system has mainly three main elements; information source which must be delivered or transmitted to a particular destination over a channel. The channel contains random noise, which adds to or otherwise disturbs the signal in some way [5].

Shifting from normal to excellent maintenance is highly required in order to achieve maximum availability of different networks since the list of applications involving the use of communication in one way or another is almost endless [6].

Performing normal maintenance is not enough to keep these networks sustainable due to the huge number of assets, variety number of applications, and developed technologies involved. This is besides that communication networks plays a major role in controlling and monitoring operations and managing crisis within and among the main sectors in power utilities (Generation, Transmission & Distribution).

In addition to the above, introducing the concept of smart grid with different applications such as smart meters ensures the requirement of excellent maintenance to communication systems since these systems play a pivotal role in transmitting metering data, publishing price information, and sending grid protections and control commands [7].

The issues of security which requires a real time communication for all devices in order to secure and monitor any expected threats makes the requirement of excellent maintenance very important.

2 Excellent at Different Stages

In order to achieve excellent maintenance, the focus should not be only on the maintenance stage only. All stages including design, engineering and commissioning should be targeted.

2.1 Excellent at Design Stage of the Equipment

The first stage is to ensure excellence during the design of the equipment. Focus at this stage should be on technical specifications and requirements for the selection of the equipment such as equipment size, panel, security, supply and interfaces.

As an example, the following requirements can be considered while ensuring the excellence of design:

- Sufficient number of equipment interfaces and slots.
- Availability of the required and essential equipment interfaces.
- Equipment Interfaces arrangement.
- Easiness of equipment interfaces installing and removing.
- The redundancy of equipment interfaces.
- Availability of cooling fans if required.
- Arrangement of equipment cables and accessories.
- Future Expansions.

Another aspect of the design is the software part. Focus should be on the following points:

- Availability of the required and future features for the equipment software.
- The easiness and friendly used of the equipment configuration.
- The possibilities of software updates.

An important aspect in the design stage is the network management system (NMS). Network Management System is the processes and techniques utilized to monitor and control a communication network, including its configuration, allocation of resources, error management, and security “Ref. [8]”. The NMS will provide an important factor for excellence. Many operations, trouble shootings and configurations are done through NMS, so the following points to be considered while preparing the specification for the NMS:

- Friendly used in terms of platforms and commands.
- Fulfilling security standards.
- Possibilities of updates and upgrades.
- Possibilities of future technologies such Artificial Intelligence.
- Availability of standby concepts.

2.2 Excellent at Engineering Stage of the Equipment

The second stage is to ensure excellence during engineering stage. Focus in this stage should be on three main aspects:

- Fulfilling the requirement with reference to the specifications.
- Excellence of the equipment supplier.
- Testing the equipment in factory or lab with excellent testing equipment and methods.

This stage is important for excellence since it is an advanced stage before inserting the equipment in the live network where any modifications will be more difficult.

2.3 Excellent at Commissioning Stage of the Equipment

The main objective of commissioning is to ensure the main functions of the equipment before inserting to existing telecommunication network. Testing the equipment should be excellent. This will be achieved by four main points:

- Sufficient and calibrated test equipment.
- Qualified testing engineers.
- Well written testing method statement.
- Passing all required tests.

In order to ensure excellence during commissioning, three main points to be considered:

- Clear test sheet including all required points that ensure the excellent operation of the equipment.
- Clear test sheet for integrating and operating the equipment in network management system.
- Checklist including supporting points that can affect the operation.

This checklist will include the following points:

- Ensure equipment fan operation and cleanness.
- Ensure room and panel temperature is sufficient.
- Ensure the healthiness of the equipment power source.
- Ensure the room and the panel are clean.
- Ensure all expected dust sources are cleared.

2.4 Excellent at Maintenance Stage of the Equipment

All above stages are mainly serving the last stage, which is maintenance since it will last for longest period. Ensuring excellence at previous stages will support and ensure the maximum level of excellence during maintenance stage; however, it is required also to focus on certain points during this stage to ensure the continuity of excellence until the end life of the equipment, besides that effective maintenance reduces overall company cost [9].

After inserting the equipment in the network, maintenance will be the main task for engineers. Many engineers focus during maintenance on the operational factors which are very important; however, this focus should be followed by other factors which might affect the operation such as temperature of the room and panels, dust and holes inside the panel, labeling and availability of panel keys.

As an example, Fig. 1 shows the effect of dust on fans, which is part of communication equipment. Failure of equipment fans may lead to high temperature, which may lead to some component damages. This in addition to the cost of more spares and additional efforts done by engineers.

Fig. 1. Examples of dust inside communication equipment and on equipment fan

There are several types of maintenance that are followed to ensure the excellence of the equipment.

The first type is preventive maintenance which is the regular maintenance carried to ensure the operation and availability of the equipment. It is the fundamental, planned maintenance activity designed to improve equipment life and avoid any unplanned maintenance activity [10].

Another maintenance type is close to preventive is Proactive maintenance which means to act before breakdown occur. It recognizes and addresses situations to prevent them from ever becoming urgent problems and breakdowns [11]. Although some might put it as another type, I prefer to link it to preventive maintenance since they have almost the same objectives.

These preventive maintenance activities will be carried based on clear plans.

The periodic of these plans will depend on the criticality of the equipment.

Following a checklist will support achieving excellent preventative maintenance taking into consideration that checking shall be conducted at site and in the network management system. The site checklist will include the following main points:

- Healthiness of all equipment interfaces.
- Healthiness of all applications passing through the equipment.
- Healthiness of all configurations.
- Non-existence of alarms.
- Cleanness of the Equipment.
- Healthiness of the Panel light.
- Healthiness of equipment Power Supply.
- Clarity and healthiness of equipment labels.
- Confirming security measures availability on the equipment.

The maintenance that is carried out through network management system will have a checklist, which will include the following main points:

- Non-existence of alarms on interfaces.
- Recording of historical alarms if exist.
- Optical measurements.
- Equipment clock priorities.
- Configuration backups.
- Healthiness of all configurations.

The second type is the corrective maintenance. This type of maintenance will focus on rectifying and solving problems occurring on the equipment. It is the performance of unplanned maintenance tasks to restore the functional capabilities of failed or multifunctioning equipment or systems [12].

Types of problems can be:

- Existence of alarms on the equipment interfaces.
- Non-communicating of applications passing through the equipment.
- Failures of equipment fans.
- Failure of equipment synchronization.
- Failures of panel light.

In order to ensure the excellence during corrective maintenance, it is necessary to establish and follow a procedure for each type of problem.

This procedure will contain the following main points:

- Responsibilities for each concern party.
- Steps to be carried out to solve the problems.
- Tools required for carrying out the process.
- Communication methods used during solving the problem.
- Documentation and recording.

The third type of maintenance is related with old equipment that is decommissioning process. Decommissioning means taking out the equipment from service. This is done mainly with old equipment that has no support from suppliers and unavailable spares in the market. While any equipment is decommissioned, all applications passing through it supposed to be shifted to another equipment.

In order to achieve excellence during decommissioning process, it is necessary to have a clear and approved procedure to be followed. This procedure will contain the following points:

- Identify all existing applications on the equipment planned to be decommissioned.
- Arrange all required tools.
- Identify and arrange all required hardware requirement.
- Shift all applications to another equipment and confirm healthiness.
- Disable all alarms for the decommissioned equipment and remove all related cables.
- Remove the equipment from site and shift to store and scrap it.
- Document and record all activities done during the process.

3 Excellence and Equipment Age

All equipment have life cycle starting from design stage up to decommissioning or replacement stages. Manufactures are stating MTTF (Mean time to failure) which is for a stated period in the life of an item the ratio of cumulative time to the total number of failures in equipment [13] in technical sheets that will indicate the life of the equipment/interfaces and usually it will be long.

Following excellent in design, engineering, commissioning and maintenance stages will provide longer life for the equipment. The same concept applies to our cars. The better maintenance we perform, the better and longer the car will stay with us.

It is necessary to follow all above points in all stages since ignoring any point during any stage might affect in later stages which can cause cost and efforts to the utilities.

4 Supporting Factors for Excellent Maintenance

There are several factors that supports excellent maintenance. Although they are not part of core maintenance process, they are essential to achieve excellence. These supporting factors are:

1. Safety: It is important for maintenance engineers to follow all safety instructions and guidelines during maintenance activities. This will ensure protection of engineer's lives and equipment operations.
2. Communication: Effective communication among engineers and managers will support excellent maintenance. This effectiveness will include the ways and time of communication for any observations, changes and activities done.
   Existence of communication strategy or process will support the effectiveness also.
3. Systems: Existence of systems will support excellent maintenance in terms of planning the activities, assigning resources, recording the activities and attaching required document.
   All of these points are important for maintenance engineers since it will save time and efforts in many aspects such as planning and recording.
4. Transportation: Maintenance engineers need to move to different locations. This requires availability of sufficient transportations.
   All transportation types should be in good and safe conditions in order to provide safety to engineers.
5. Tools: All tools required to perform maintenance activities should be available with engineers. These tools differ from one field to another. It is important to maintain these tools and keep them in safe and secure places.
6. Training: Since technologies, processes, and systems are continuously developing, training is required to perform maintenance at excellent level.
   These trainings should be periodic and it should be ensured that all engineers gathered and understood the required knowledge in order to implement well during maintenance.
7. Knowledge Sharing: There are many issues and situations faced maintenance activities. All of these should be shared among concern engineer since this will benefit all.
   Having the required knowledge from a similar issue or situation faced by other will ease the maintenance process especially during problems.

5 Case Study (Implementation of Equipment Excellence Model)

In Dubai electricity and Water Authority (DEWA) - telecommunication department, a project was applied called (X-star communication panel) where a total number of 151 equipment were evaluated and rated. The objective of this project was to evaluate the excellence of communication equipment inside substations and rate them. This was to improve the existing maintenance process and specifications of future equipment.

The project was to evaluate each panel according to 15 criteria. Each criteria will have different weight. Accordingly, the number of stars will result and it will be placed on the equipment as shown in Fig. 2.

Fig. 2. Examples of stars on communication equipment

Criteria selected were focusing on different factors within the communication rooms and panels within transmission power substations where these rooms and panels exist.

There were 15 criteria selected to carry out this project were as follows:

1. Cleanness of the panel: This criterion will focus mainly on the panel dust level. It is important that the panel to be clean for the following reasons:

   - To protect the assets inside the panel.
   - To provide suitable environment for maintenance engineers.
   - To show labels, information, or plates displayed on the panel.

2. Availability of Panel keys: All panels are closed in normal conditions with locks. Special keys are needed to open and close these panels.
   These keys are available either in the room itself, in common area in the substation or carried with engineers.
   This criterion will check whether the right key is available in its desired location with healthy condition in order to provide easiness for engineers during maintenance.
3. Panel Plate: There are many panels within the communication room, so it is important that each panel to have its right plate. This to reduce confusions during maintenance activities and avoid opening wrong panels.
   These plates should have clean and understandable writings to reflect the actual asset.
4. Door condition of the panel: This criterion will focus on outside and inside doors of the panel whether they are opening and closing properly.
   It will ensure also that keys are opening and closing these doors. In addition, it will confirm that the door is not touching any asset inside the panel.

5. Extra Equipment in the panel: This criterion will ensure that there are no extra equipment either decommissioned or not in use inside the panel.
   These extra equipment will take space in the panel, so it could cause difficulties during maintenance especially if the equipment is big.
   The normal condition is that all decommissioned or not equipment to be removed from the panel and delivered to store or scarped.
6. Cleanness of the Equipment(s): Another level of cleanness is ensured by checking equipment cleanness. All equipment should be dust free or have an acceptable dust level which will not affect the equipment.
   This criterion is important since dust can damage the equipment or affect the operation of it.
7. Panel Light: Each panel has its own light that will provide visibility inside it. Light position and type will differ among the panels.
   This criterion will ensure that the light is working beside checking light switches if available.
8. Noisy Sounds: Some equipment will generate sounds. Mainly in communication equipment, this can be due to equipment fans. It is necessary that these sound to be acceptable in order to maintain healthy environment during maintenance.
9. Alarms: This criterion will ensure that there are no alarms existing on the equipment interfaces.
10. Drawings: It is necessary to have a reference for the panel and equipment layout and details. This will include the equipment installed in the panel, cable arrangements, MCB (Miniature Circuit Breaker) details, alarms connections…etc.
    Existence of the drawings in the panel will support the maintenance engineers in case required any information related to panel or equipment.
    Beside the existence of the drawings, it is necessary to make sure that they are the approved ones and with healthy condition.
11. Additional cable: This criterion will ensure that there are no additional cables laid inside the panel since this will cause difficulty for the maintenance engineers while working in the panel.
12. Holes: Holes are existing in the panels to run the cables and fiber patch cords. If they are kept open, they will be a main cause for dust entrance into the panel, so it is necessary that all holes are covered and closed well.
13. Organization of Cables: All cables within the panel should be organized or properly looped in a suitable location in the panel. This is to provide the engineer a comfortable area to work inside the panel.
14. Loose Cables: This criterion will ensure that all cables are connected to their respective ports, so no loose cables exit in the panel.
    Loose cables may effect engineers operation's inside the panel.
15. Labels: This criterion will confirm that labels are existing on each equipment, cables and ports.
    It is necessary to check the following related to labels:

    - Healthiness of the labels.
    - Right descriptions for each label.
    - Clarity of the label.

Each criterion will be given a weight based on its importance and criticality to the operation and maintenance of the equipment.

Table 1 shows the criteria and the weights used to evaluate the excellence level of the equipment.

Table 1. Criteria and weights

No	Criteria	Weight
1	Cleanness of the panels	5%
2	Availability of Panel keys	3%
3	Plate showing the name of the equipment placed on the front of the panel	2%
4	Outside & Inside door condition of the panel	5%
5	No Extra Equipment in the panel (Decommissioned or not in use)	3%
6	Cleanness of the Equipment(s)	7%
7	Condition of Panel inside light	5%
8	No noisy sounds coming from equipment(s)	10%
9	No physical alarm on the equipment panel	10%
10	Equipment drawing available in the panel	5%
11	No additional cable/patch cords length looped inside the panel	10%
12	All unwanted holes in the panel are closed	10%
13	Cables/patch cords are organized in the panel	10%
14	All cables are connected to their respective ports	5%
15	Availability and condition of Labels	10%

The evaluation process will be conducted by two mangers where they will evaluate each criteria and update the information in a table to get the final rating for the panel.

Table 2 shows an example of evaluating an existing communication equipment carried out by two managers, who evaluated each criteria and accordingly, the final result (Number of Stars) will be generated.

All the observations will be communicated to the engineers in order to take the required action since the objective is to the enhance the maintenance process.

There were several outcomes from this project:

- Identify hidden risks that can affect the equipment (Example: unavailability of panel light may raise safety issues since the panel will be dark from inside).
- Equipment which got five stars were set as a 'model'.
- The most important lesson is that the excellence of the equipment will depend on all stages.

Table 2. Examples of evaluation and ratings

Criteria	Excellent = 10, Average = 5, Poor = 0		Average	Weight	Point
	Manager 1	Manager 2			
Cleanness of the panels	**0**	**5**	2.5	5%	0.125
Availability of Panel keys	**10**	**10**	10	3%	0.3
Plate showing the name of the equipment placed on the front of the panel	**5**	**0**	2.5	2%	0.05
Outside & Inside door condition of the panel	**10**	**10**	10	5%	0.5
No Extra Equipment in the panel (Decommission or not in use)	**10**	**10**	10	3%	0.3
Cleanness of the Equipment(s)	**5**	**5**	5	7%	0.35
Condition of Panel inside light	**0**	**0**	0	5%	0
No noisy sounds coming from equipment(s)	**10**	**10**	10	10%	1
No Physical alarm on the equipment panel	**5**	**10**	7.5	10%	0.75
Equipment drawing available in the panel	**10**	**10**	10	5%	0.5
No additional Cable/Patch cords length looped inside the panel	**5**	**5**	5	10%	0.5
All unwanted holes in the panel are closed	**10**	**10**	10	10%	1
Cables/patch cords are organized in the panel	**10**	**5**	7.5	10%	0.75
All cables are connected to their respective ports	**5**	**5**	5	5%	0.25
Availability and condition of Labels	**10**	**10**	10	10%	1
Total points based on Weighted	**73**	**74.5**		**100%**	**73.75**
Total Starts	**3**	**3**			**3**

6 Conclusion

In conclusion, the following recommendations can be listed:

1) Excellence is important subject for any utility and it should be applied in communication field.
2) Focus on excellence culture among engineers.
3) Focus on all stages in order to achieve longer life for communication equipment.
4) Neglecting any criteria in any stage will affect the overall excellence of the equipment.

5) All excellence criteria can be considered as hidden risks if we didn't focus on them.
6) Availability of high standard specifications and checklists are important for all stages.
7) Mock drills on telecom equipment's and feedback reports from service owner are important tools to improve equipment excellence.

References

1. Sutton, D.: A Practitioner's Guide to Operations Excellence, 2nd edn., p. XI. Operations Excellence Svcs LLC (2012)
2. Capehart, B.L., Turner, W.C., Kennedy, W.J.: Guide to Energy Management, 7th edn., chap. 10, p. 363. The Fairmont Press, Inc. (2011)
3. Tomasi, W.: Electronic Communications Systems. Fundamentals through Advanced, 3rd edn., chap. 1, p 1. Pearson (1997)
4. Haykin, S.: Communication Systems, 3rd edn., chap. 1, p. 19. Hamilton Printing Company (1978)
5. Stanley, W.D.: Electronic Communication Systems, chap. 1, p. 1. Prentice-Hall, Inc. (1982)
6. Haykin, S.: Communication Systems, 3rd edn., chap. 1, p. 1. Hamilton Printing Company (1978)
7. Bakken, D.: Smart Grids, 1st edn., chap. 11, p. 254. Taylor & Francis Group (2014)
8. Kplan, S.M.: Wiley Electrical and Electronics Engineering Dictionary, 1st edn., p. 156. Wiley (2004)
9. Palmer, D.: Maintenance Planning and Scheduling, 3rd edn., chap. 1, p. 3. The McGraw-Hill Companies (2013)
10. Wireman, T.: Preventive Maintenance, 1st edn., chap.1, p. 1, Reliabilityweb.com (2008)
11. Palmer, D.: Maintenance Planning and Scheduling, 3rd edn., chap. 1, p. 2. The McGraw-Hill Companies (2013)
12. Smith, A.M., Hinchcliffe, G.R.: RCM. Gateway to World Class Maintenance, 1st edn., chap. 2, p. 20. Elsevier Inc., (2004)
13. Smith, D.J.: Reliability Maintainability and Risk, 7th edn., chap. 2, p. 14. Elsevier Butterworth-Heinemann (2005)

Development of a Two-Level Output Vehicle Safety Device Initiated by the Driver's Eye Movement Patterns

Efren Victor Jr. N. Tolentino(✉), Joseph D. Retumban, Dann Adrian A. Dimalibot, Mark Leo S. German, Nelson Lee L. Martinez, Edwin C. Pangatungan, and Carl Louise M. Quintero

National University, Philippines, 1008 Manila, Philippines
evntolentinojr@national-u.edu.ph

Abstract. Driver's drowsiness is still one of the major causes of vehicular-related incidents especially in the Philippines. In 2019, a 4.16% increase in road accidents was tallied and the most dangerous hour was singled out between 1:00AM to 2:00AM. In this study, a low-cost two-level output vehicle safety device was developed to be installed inside a car in order to deter road accidents caused by driver's drowsiness. The detection of the driver's drowsiness level was based on eye blink count exceeding 500 ms in one-minute duration. When 5 counts of eye blink exceeding 500 ms in one minute were recorded, a Level 1 output will be triggered reminding the driver to stay awake during the drive. On the other hand, when 10 counts of eye blink were recorded, Level 2 output will be triggered suggesting to the driver to take some rest and signaling other surrounding vehicles that something is wrong with the car. OpenCV-Python was used as the programming language and Raspberry Pi for eye blink detection. By synchronizing the process of the programming language, the output was considered a success.

Keywords: Drowsiness detection · Eye movement · Vehicle safety device

1 Introduction

Road accidents are still a major concern of many developed and developing countries despite the efforts in improving their infrastructure and transportation systems [1]. In the Philippines, countless vehicular-related incidents causing fatalities were recorded and the numbers of these incidents are appallingly increasing. In Manila alone, the Metropolitan Manila Development Authority (MMDA), through the Metro Manila Accident Reporting and Analysis System (MMARAS), counted a total of 121,771 road incidents in 2019, a 4.16%-increase from the previous year. Of these numbers, cars have the highest accident involvement and that the most dangerous hour is from 1:00AM to 2:00AM probably because "motorists take advantage of wider roads, are tired, or are under the influence of alcohol while driving" [2].

Driver's tiredness, as cited as one of the most common causes of road accidents, can be directly linked to driver's drowsiness during the drive. However, drowsy driving

M. Ma (Ed.): ICTCE 2020, LNEE 797, pp. 135–142, 2022.
https://doi.org/10.1007/978-981-16-5692-7_14

2 Methodology

2.1 Project Development

The vehicle safety device consists of three level outputs – the sensor for detecting the drowsiness level of the driver, the vibration pads, and the alarm which will serve as a reminder to the driver to stay awake during the drive. Figure 1 below gives a rough sketch of the prototype.

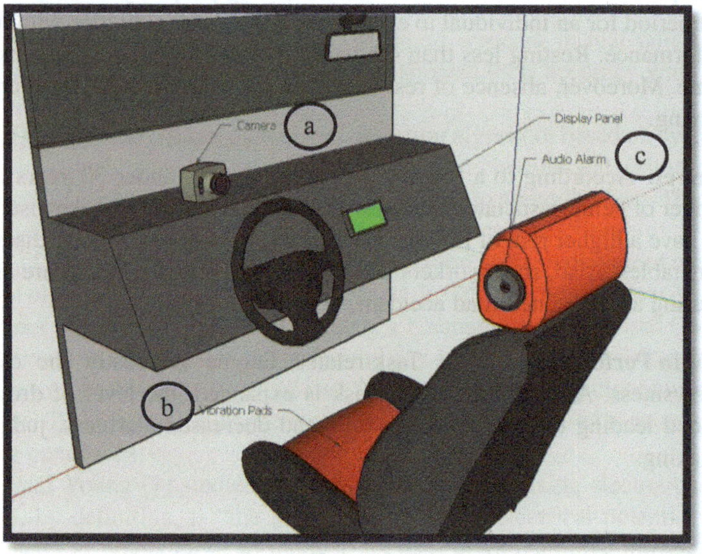

Fig. 1. Sketch of the prototype showing (a) Camera for eyelid movement detection, (b) Vibration pads, and (c) Audio alarm.

The detection of the level of drowsiness was done using a sensor that can detect the movement sequence of a person's eyelids via a 5 megapixel camera. The data gathered were sent to a Raspberry Pi which uses OpenCV-Python to determine the action to be taken by the device. For Mild drowsiness detection (Output Level 1), the system will try to awaken the driver by triggering an alarm placed on the car's headrest. The display LCD will also be activated to remind the driver that he is sleepy and needs to be careful in driving. On the other hand, for Extreme drowsiness detection (Output Level 2), the system will trigger not just the alarm but also the vibration pads placed on the car seat and the hazard light that will notify the surrounding vehicles that something is wrong with the car. Figure 2 shows the schematic diagram of the drowsiness detection system.

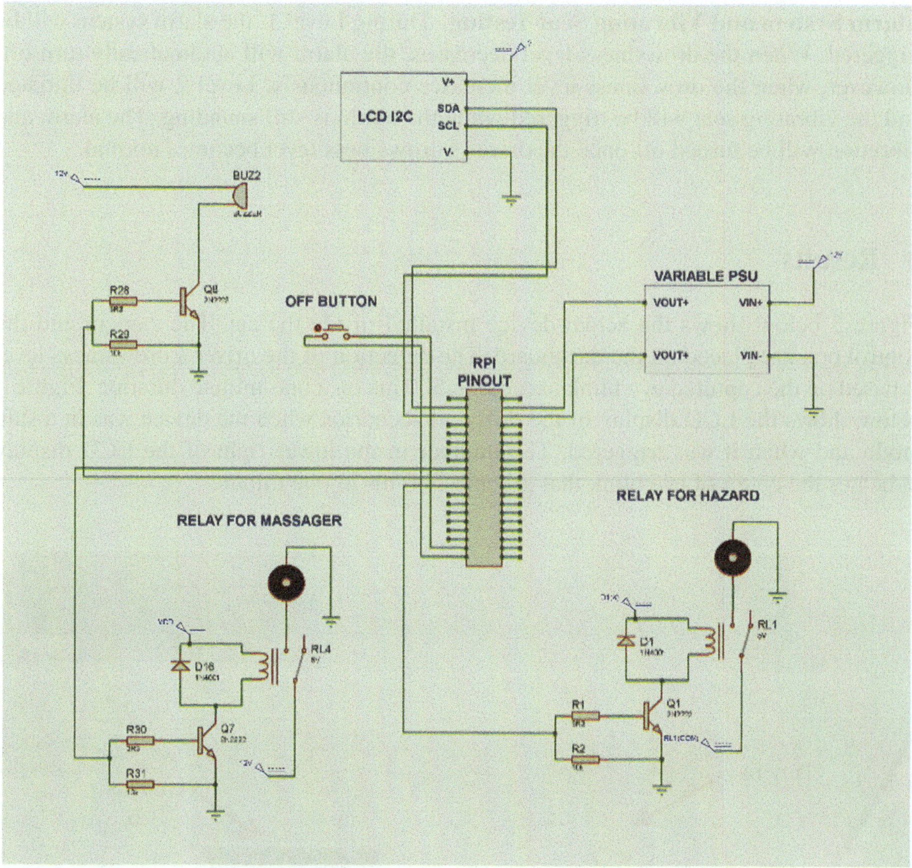

Fig. 2. Drowsiness Detection System schematic diagram.

2.2 Operation Testing and Procedure

Drowsiness Level Detection System. The eyelid movement of the driver will be the input signal of the camera. The camera will capture and count the driver's blink that exceeds 500ms every one minute. By definition, an eye blink is an eyelid closure with a duration of 50 to 500 ms (ms) and closures in excess of 500 ms are considered as microsleep episodes [12].

The data gathered will be sorted out accordingly. For Mild drowsiness detection (Output Level 1) to be triggered, the system should capture 5 blinks exceeding 500 ms for one minute duration. On the other hand, for Extreme drowsiness detection (Output Level 2), the system should capture 10 blinks for one minute duration.

LCD for Preemptive Notification and Hazard Light Testing. During Level 1, a message will pop out of the screen saying "Maintain speed at 50 kph maximum" while during Level 2, the pop out message is "Stop the car! Take a nap". Moreover, during Level 2, the hazard light will be activated.

Alarm System and Vibrating Seat Testing. During Level 1, the alarm system will be triggered. When the drowsiness level decreases, the alarm will automatically turn off. However, when the drowsiness level increases continuously, Level 2 will be initiated and the vibrating seat will be triggered while the alarm is still sounding. The alarm and vibration will be turned off once the driver's drowsiness level becomes normal.

3 Results

Figure 3 below shows the actual device installed inside the car. The camera and the control box are placed on the dashboard. The detection of the driver's drowsiness level is based on the captured eye blink exceeding 500 ms on a one-minute duration. Figure 4 below shows the LCD display of the different scenarios when the device was in a safe mode and when it was triggered. The number in the lower right of the LCD display indicates the count of eye blink that exceeds 500 ms in one minute.

Fig. 3. The actual device showing the camera for eye movement detection and the control box.

Upon testing the device, the Raspberry Pi attached in the control box detects the movement of the driver's eyelid patterns. Based on the counting principle of Raspberry Pi, one duration is equal to one millisecond. When the blinking duration of the driver reached 500 ms (or 5 durations in Raspberry Pi), the device will record a count of 1. When

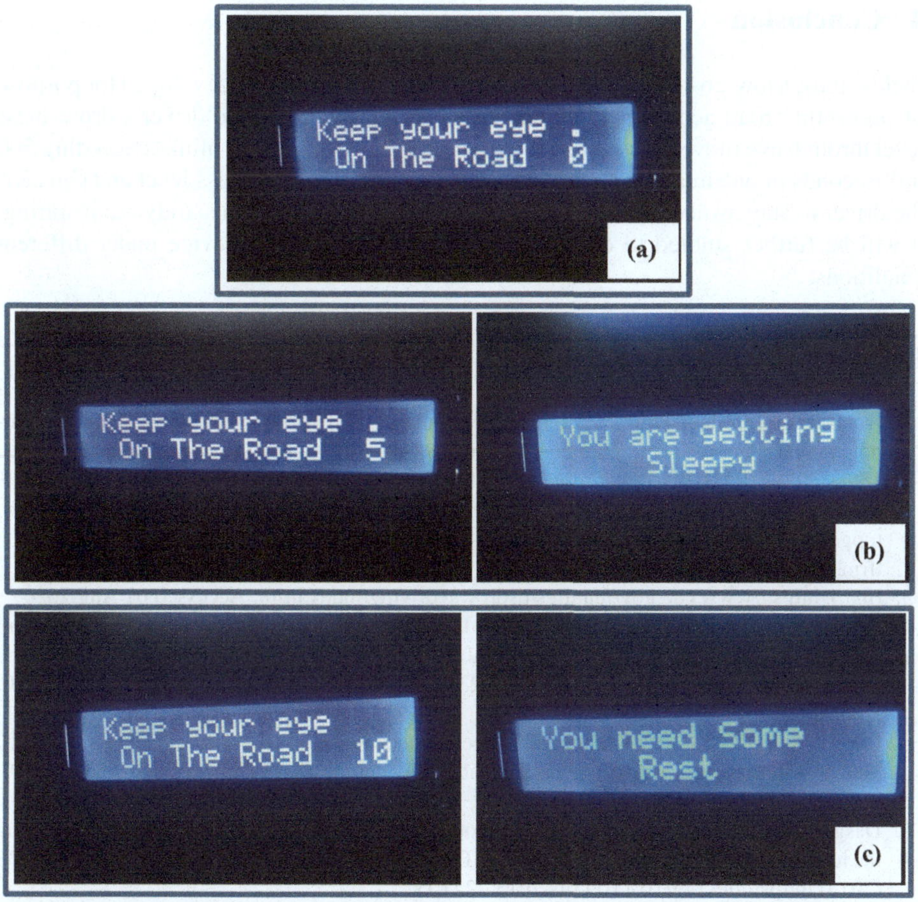

Fig. 4. LCD display for different scenarios: (a) safe condition where the driver is not at all sleepy, (b) Output Level 1 is triggered, and (c) Output Level 2 is triggered.

the blinking pattern continues until 5 counts, Level 1 will be activated. The sound alarm will be triggered reminding the driver to stay awake. When the driver is still drowsy and the count becomes 10, Level 2 will be activated. Aside from the sound alarm, vibrating pads located on the car seat will be triggered. Another safety measure that is triggered during Level 2 is the hazard light.

4 Conclusion

In this study, a low-cost two-level output vehicle safety device was developed for purpose of preventing road accidents. The device works by detecting the driver's drowsiness level through eye movement patterns. By counting the driver's eye blinks exceeding 500 milliseconds in one minute, the device can determine the drowsiness level and can alert the driver to stay awake during the drive. Though the results of the study are inspiring, it will be further studied to establish the predictability of the device under different conditions.

References

1. Doudou, M., Bouabdallah, A., Berge-Cherfaoui, V.: Driver drowsiness measurement technologies: current research, market solutions, and challenges. Int. J. Intell. Transp. Syst. Res. **18**(2), 297–319 (2019). https://doi.org/10.1007/s13177-019-00199-w
2. ICYMI: Metro Manila road-accident stats for 2019, LRT-2 repair update. (n.d.). https://www.topgear.com.ph/news/motoring-news/news-roundup-29-february-2020-a2619-20200229-lfrm. Accessed 01 Aug 2020
3. (n.d.). https://www.cdc.gov/features/dsdrowsydriving/index.html. Accessed 01 Aug 2020
4. Danisman, T., Bilasco, I.M., Djeraba, C., Ihaddadene, N.: Drowsy driver detection system using eye blink patterns. In: International Conference on Machine and Web Intelligence (ICMWI 2010), Alger, Algeria, pp. 230–233 (2010). https://doi.org/10.1109/ICMWI.2010.5648121
5. Liu, D., Sun, P., Xiao, Y., Yin, Y.: Drowsiness detection based on eyelid movement. In: 2010 Second International Workshop on Education Technology and Computer Science (2010). https://doi.org/10.1109/etcs.2010.292
6. Dasgupta, A., George, A., Happy, S.L., Routray, A.: A vision-based system for monitoring the loss of attention in automotive drivers. IEEE Trans. Intell. Transp. Syst. **14**(4), 1825–1838 (2013). https://doi.org/10.1109/tits.2013.2271052
7. Ebrahim, P., Stolzmann, W., Yang, B.: Eye movement detection for assessing driver drowsiness by electrooculography. In: 2013 IEEE International Conference on Systems, Man, and Cybernetics (2013). https://doi.org/10.1109/smc.2013.706
8. Hussein, W., El-Seoud, M.S.: Improved driver drowsiness detection model using relevant eye image's features. In: 2017 European Conference on Electrical Engineering and Computer Science (EECS) (2017). https://doi.org/10.1109/eecs.2017.55
9. Horne, J., Reyner, L.: Vehicle accidents related to sleep: a review. Occup. Environ. Med. **56**(5), 289–294 (1999)
10. Dinges, D.: An overview of sleepiness and accidents. J. Sleep Res. **4**, 4–14 (1995)
11. Skipper, J.H., Wierwille, W.W.: Drowsy driver detection using discriminant analysis. Human Factors: J. Human Factors Ergon. Soc. **28**(5), 527–540 (1986). https://doi.org/10.1177/001872088602800503
12. Wang, Y., Toor, S.S., Gautam, R., Henson, D.B.: Blink Frequency and duration during perimetry and their relationship to test-retest threshold variability. Invest. Opthalmol. Vis. Sci. **52**(7), 4546 (2011). https://doi.org/10.1167/iovs.10-6553

Design and Implementation of V-Band MMIC Low Noise Amplifier in GaAs mHEMTs Process

Wentao Zhu, Debin Hou, Jixin Chen, and Wei Hong(✉)

State Key Laboratory of Millimeter Waves, School of Information Science and Engineering, Southeast University, Nanjing 210096, People's Republic of China

Abstract. In this paper the design and implementation of a MMIC low noise amplifier (LNA) operating at 50–75 GHz band in 70 nm GaAs mHEMTs process is presented. The measured minimum noise figure (NF) is 2.7 dB and the maximum gain is 19.3 dB. The LNA is composed of three stages in cascaded common-source topology with more than 16.4 dB gain and less than 3 dB NF over the whole band. It is applicable in millimeter wave (mmWave) communications and measurement instruments.

Keyword: GaAs · Broadband · LNA · V-band

1 Introduction

In recent years, mmWave have been widely used in 5G communications, radar, and remote sensing etc. with the characteristics of extremely wide bandwidth, narrow beam, and smaller device size. A major breakthroughs have been made in the development of mmWave transmitters, receivers, antennas, and millimeter wave devices. For a communication system, the receiver needs to amplify the signal when receiving it from the antenna, and the amplifier itself also has certain noises, which will seriously affect the performance of the system. Therefore, the receiver always needs a LNA to meet the requirement of the signal to noise ratio (SNR).

Monolithic Microwave Integrated Circuit (MMIC) is often used as the functional circuits in microwave and millimeter wave modules. The process used in this paper is the D007IH process provided by OMMIC, which is an advanced GaAs process. It has an industry-leading 70 nm dual mushroom structure gate, and the channel indium concentration is as high as 70%. Its cut-off frequency Ft = 300 GHz and the maximum frequency Fmax = 450 GHz. Because of its excellent noise performance, the products made by OMMIC using D007IH technology have a very low noise figure, up to 2.8 dB (typical value) @90 GHz, which are suitable for communication, imaging and radar applications.

The V-band (50 GHz–75 GHz) has attracted many attention because of including the 7 GHz unlicensed bandwidth (57.2 GHz–65.8 GHz). This frequency band is mainly used for short-distance wireless communications and inter-satellite communications etc.

M. Ma (Ed.): ICTCE 2020, LNEE 797, pp. 143–147, 2022.
https://doi.org/10.1007/978-981-16-5692-7_15

2 Circuit Design and Simulation

To design a LNA, it is necessary to choose the size of transistor firstly. Then design the bias and impedance matching circuits around the transistor. Finally, the design of layout and electromagnetic field full-wave simulation.

Considering the number of gate fingers is no more than 8, the width of a single gate finger satisfies 12.5 μm ≤ Wu ≤ 75 μm. Then the transistors with different gate fingers are simulated for comparing the maximum gain (MaxGain) and minimum noise figure (NFmin). Selecting the small signal model in the library file firstly, then set the gate-source voltage Vgs to −0.1 V and set the drain-source voltage Vds to 1.0 V. The simulation results are shown in Fig. 1.

The results show that when choosing 6 or 8 gate fingers, the maximum gain of transistors with a width of each gate finger greater than 30 μm decreases with increasing frequency. The influence of parasitic capacitance becomes greater and the input impedance is close to the short-circuit state.

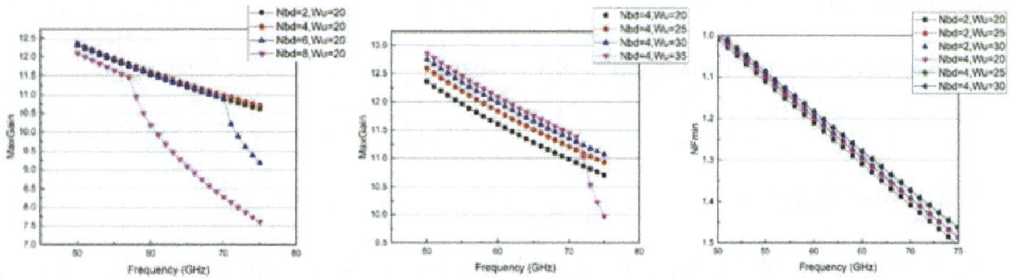

(a) (b) (c)

Fig. 1. (a) MaxGain with different Nbd. (b) MaxGain with different Wu. (c) Nfmin of transistors with different sizes

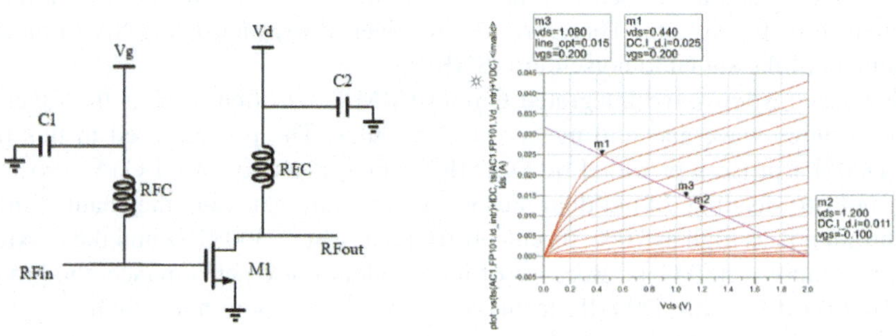

(a) (b)

Fig. 2. (a) Common structure of bias circuit. (b) DC-simulation results of transistors.

In order to obtain higher gain, it can be seen from Fig. 1 that when Nbd is 4, as Wu increases, it is more susceptible to parasitic capacitance at high frequencies. Therefore,

It is more suitable to choose transistors with Nbd of 2. Under the same voltage as above, when Nbd is 2, 4, and Wu is 20, 25, 30, the noise situation is shown in Fig. 1 (c).

Therefore, considering both gain and noise matching, it is better to choose a transistor with a size of Nbd = 2 and Wu = 25μm for the low-noise amplifier designed.

The bias network is designed which can not only provide an appropriate static operating point for the active device under specific operating conditions but also suppress the discreteness of transistor parameters and the influence of temperature changes to maintain constant operating characteristics. In order to make the field effect transistor work in the linear region, it is necessary to provide an appropriate voltage to the gate and drain of the transistor for the common-source amplifier structure. The key point is that the bias condition of the FET usually requires a negative gate voltage. The bias structure commonly used in transistors is shown in Fig. 2.

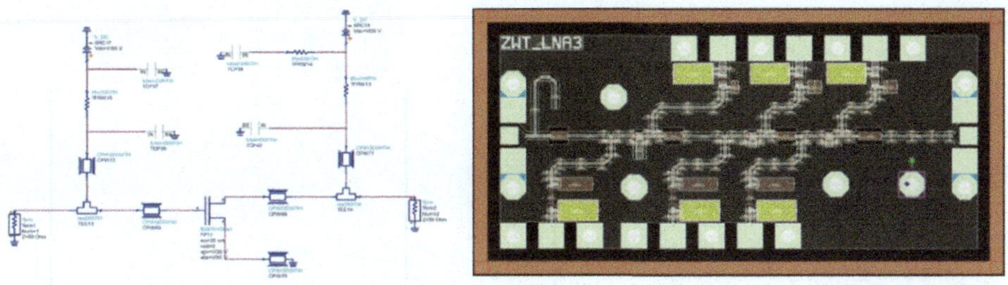

(a) (b)

Fig. 3. (a) Circuit of single-stage LNA. (b) The layout of whole LNA.

The bias network requires two power supplies with different polarities. In practical applications, the isolation and the gain of this structure is worse. This design slightly changes the structure by adding a typical RLC structure to improve the isolation.

According to the DC-simulation results, the drain current should be around 10mA. The m2 point in the Fig. 2(b) is the selected DC bias point, where vds is 1.2 V, vgs is −0.1 V and Id is 11mA. The gate and drain voltages need to be provided by two different sources.

The first-stage circuit of the designed LNA is shown in Fig. 3, then the three-stage LNA is designed with the same idea. The inter-stage impedance and input and output impedance are matched, and the layout design is completed by electromagnetic field simulation. During the electromagnetic simulation, the electromagnetic field conditions at the similar GCPWs are emulated to avoid coupling between the lines that are too close to affect the performance of the LNA. The layout of the designed three-stage cascaded LNA is shown in Fig. 3 (b).

According to the layout designed, use sonnet to simulate the electromagnetic field and replace each component of the schematic with the s2p-file of the simulation result. The simulation results are shown in Fig. 4 (a), including S parameters, stability coefficient Mu value and noise figure.

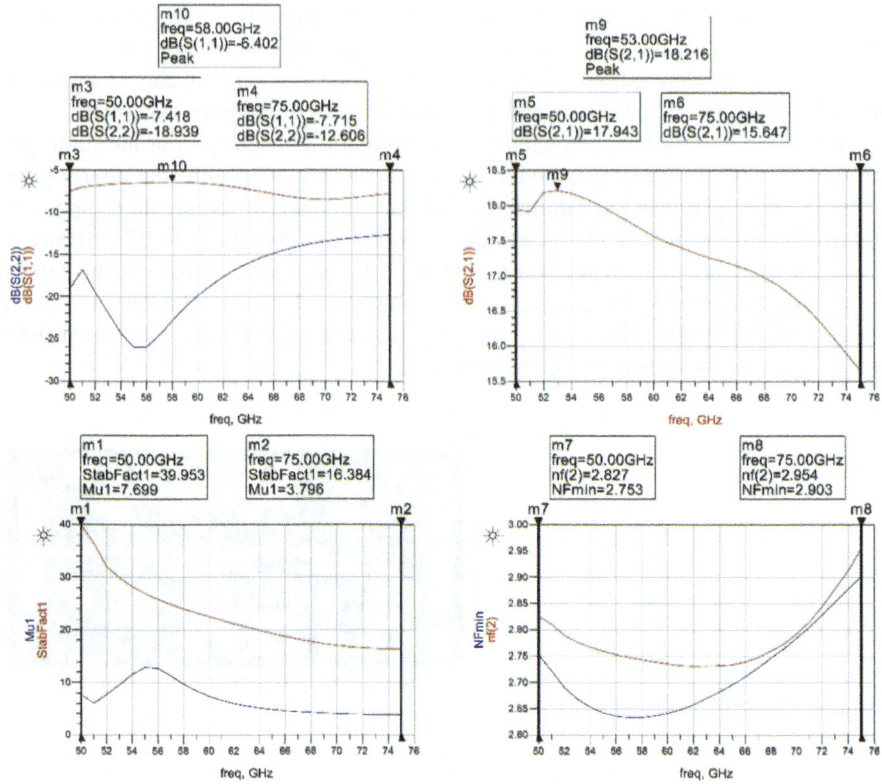

(a)

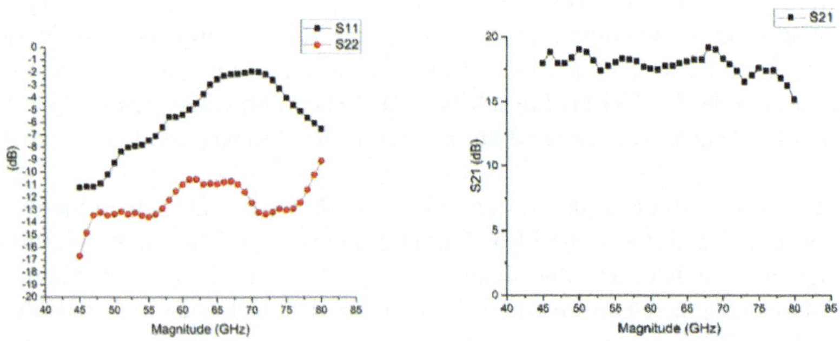

(b)

Fig. 4. (a) The simulation results. (b) The measured results.

Figure 4 (b) shows the S-parameters of the chip when tested in an environment where the voltage between the gate and the source is −0.1 V, the voltage between the drain and the source is 1.2 V, and the drain current is 36mA.

According to the comparison between the simulated and measured results of the LNA, it is found that S11 has a deviation in the chip test results, but the chip performance is stable during the test. The trend of S21 and S22 curves in the test results is similar to the simulation results.

3 Conclusion

A broadband MMIC LNA working at V-band (50–75 GHz) is designed and implemented in this paper in 70 nm GaAs mHEMTs process. The bias network and impedance matching network are optimally designed. The overall layout is simulated by electromagnetic field software. The simulation results achieve a maximum gain of 18.2 dB and a minimum noise of 2.7 dB within 50–75 GHz. The tested S11 is slightly worse, but the other results are basically in agreement with the simulation.

References

1. Hsieh, J.: Low-noise amplifier by using a signal-reuse wake-up technology. In: IET Microwaves, Antennas & Propagation, vol. 12, no. 3, pp. 287–294 (2018)
2. Wang, Y., Yu, W.: Design and simulation of a W-band broadband low noise amplifier. In: 2013 IEEE International Conference on Microwave Technology & Computational Electromagnetics, Qingdao, China, pp. 119–122 (2013)
3. Leuther, A., et al.: 70 nm low-noise metamorphic HEMT technology on 4 inch GaAs wafers. In: International Conference on Indium Phosphide and Related Materials, Santa Barbara, CA, USA, pp. 215–218 (2003)
4. Chiong, C., Chen, H., Kao, J., Wang, H., Chen, M.: 180–220 GHz MMIC amplifier using 70-nm GaAs MHEMT technology. In: 2016 IEEE International Symposium on Radio-Frequency Integration Technology (RFIT), Taipei, pp. 1–4 (2016)
5. Ciccognani, W., Giannini, F., Limiti, E., Longhi, P.E.: Full W-band high-gain LNA in mHEMT MMIC technology. In: 2008 European Microwave Integrated Circuit Conference, Amsterdam, pp. 314–317 (2008)
6. Ciccognani, W., Limiti, E., Longhi, P.E., Renvoise, M.: MMIC LNAs for radioastronomy applications using advanced industrial 70 nm metamorphic technology. In: 2009 Annual IEEE Compound Semiconductor Integrated Circuit Symposium, Greensboro, NC, pp. 1–2 (2009)
7. Fanoro, M., Olokede, S.S., Sinha, S.: Design of a low noise, low power V-band low noise amplifier in 130 nm SiGe BiCMOS process technology. In: 2017 International Semiconductor Conference (CAS), Sinaia, pp. 275–278 (2017)
8. Sarkar, M., Banerjee, P., Majumder, A.: Design of broadband MMIC low noise amplifier at W band using GaAs pHEMTs. In: 2017 International Conference on Innovations in Electronics, Signal Processing and Communication (IESC), Shillong, pp. 194–198 (2017)
9. Kumar, T.B., Ma, K., Yeo, K.S.: A 60-GHz coplanar waveguide-based bidirectional LNA in SiGe BiCMOS. IEEE Microwave Wirel. Compon. Lett. **27**(8), 742–744 (2017)
10. Völkel, M., Dietz, M., Hagelauer, A., Weigel, R., Kissinger, D.: A 60-GHz low-noise variable-gain amplifier in a 130-nm BiCMOS technology for sixport applications. In: 2017 IEEE International Symposium on Circuits and Systems (ISCAS), Baltimore, MD, pp. 1–4 (2017)
11. Sharma, R., Rajendran, J., Ramiah, H.: Broadband linear power amplifier for picocell basestation application. J. Commun. **12**(7), 419–425 (2017). https://doi.org/10.12720/jcm.12.7.419-425

Hidden Risks in Utilities Communication Panels

Seena Zarie(✉)

Dubai Electricity and Water Authority (DEWA), Dubai, UAE
seena@dewa.gov.ae

Abstract. Developed countries provide attention to risks for all their critical sectors. Utilities are one of these critical sectors. Communication network is a main element for any utility. Any major failure in this network will have huge impact on the utilities' operations. Focus should be always on all types of risks that have effects on operations. Some risks can be considered hidden meaning that they have in direct impact on the equipment's operation. Identifying these hidden risks in telecommunication panels will provide huge benefits for utilities since they will prevent failures and maintain sustainable operations. This paper aims to highlight some examples of hidden risks in utilities communication panels, their effects on the network and how to mitigate them. The study is based on several failures and challenges faced by DEWA (Dubai Electricity and Water Authority) communication engineers during maintenance and troubleshooting. The study concludes that it is essential to take care about and mitigate these hidden risks in all stages starting from design up to maintenance stages.

Keywords: Communication · Utility · Risk

1 Introduction

Risk is defined as a probability or threat of damage, injury, liability, loss, or any other negative occurrence that is caused by external or internal vulnerabilities, and that may be avoided through preventive action [1].

Utility is defined as category of companies that provide basic amenities, such as water, sewage services, electricity, dams, and natural gas [2]. Utilities pay high attention to risks among all related critical infrastructures. This can be observed through applying risk management programs.

Communication networks are transmission systems enabling information to be transmitted in analogue or digital form between various different sites by means of electromagnetic or optical signals [3]. Communication network is one of these critical infrastructures. Utilities depend on communication network for many purposes. Controlling and monitoring operations, managing crisis and security in all utilities sectors (Generation, Transmission and Distribution) are critical focus for any utility. Furthermore, utilities have started the implementation of smart grid where reliability and stability of communication network is very important. This network plays a pivotal role in transmitting metering data, publishing price information, and sending grid protections and control commands [4].

M. Ma (Ed.): ICTCE 2020, LNEE 797, pp. 148–152, 2022.
https://doi.org/10.1007/978-981-16-5692-7_16

The critically of communication network requires high attention to risks. Focus is mostly on major risks which can affect the operation of the equipment such as failures of critical interfaces within the equipment (Example: control units and power units), failure of equipment power supply, equipment configuration miss match or error, etc....

There are another type of risks which are not major as the above ones, however, they can lead to operational problems. These risks are hidden risks since engineers are not paying much attention to them like the major ones. These hidden risks occurs mostly in communication panels.

2 Examples of Communication Panel Hidden Risks

2.1 Unorganized Cables Inside the Panel

During maintenance activities, engineers check many points inside the panel such as measurements and physical checks. If the cables are not organized, engineers will face difficulty in performing maintenance tasks. More critical situations can occur if these cables are touching the door causing sharp bending or damage especially if the cable is light such as patch cords.

If the cables are laid randomly inside the panel or not terminated properly, it can cause also safety issues. Engineers working inside the panel can mistakenly pull out the cables while coming out from the panel. This can lead to operational problems for the equipment and the network.

All cables should be organized in the available cable trays with sufficient length, type and proper labels.

2.2 Wrong or Missing Labels

During commissioning and maintenance of the equipment, it is important to perform the right task. One of the ways to do this is to have right labels on the interfaces, connectors, cables, miniature circuit breakers (MCB) and any device inside the panel.

Wrong or missing labels can lead to wrong decisions such as switching off the wrong miniature circuit breaker (MCB) or removing the wrong connector or interface. This is besides wasting engineer's time during commissioning and maintenance while searching the right place for the task.

Labels should be well defined, clear, non-erasable and placed on correct locations within the panel.

2.3 Existence of Dust

Dust is existing everywhere and can't be eliminated, so it is important to minimize the effect of it to the minimum level. Dust can cause damage to different components and devices within the panel besides making the labels unclear. Dust can cause health impact also if it exists so much inside the panel.

It is important to identify the source of the dust. Panel doors, wholes inside the panel and location and operation of fans – if exist - within the panel are considered major sources for dust. It is important to mitigate the effect of these factors and have cleaning plan during maintenance activities.

2.4 Nonfunctional of Panel Lights

Panel light has important function during any commissioning, maintenance or trouble shooting activities. It will provide clear view inside the equipment. If the panel is dark from inside, it will be difficult to perform tasks and identify the labels, cables and devices which will result in human errors and may raise safety concerns. Some engineer uses external lights which is also not sufficient since one hand will be busy with holding the light and this will make tasks more difficult to be performed.

Panel lights should be always healthy and be replaced immediately in case of any problems. Also, it should be fixed in the right place within the panel for better inside panel vision.

The above risks can be minor for some engineers, however, if their effects are measured it can prove that these risks can cause a major effects on the equipment. There can be many other hidden risks within communication panels which can be identified. Focus should be on any factor which might have an effect - even if minor – on the equipment.

3 Mitigating the Hidden Risks

Conducting risk assessment and audits will support mitigating of the hidden risks, however, it can be identified and mitigated from earlier stages.

3.1 Design Stage

Design stage is an important stage to minimize the effect of the hidden risk. Excellent planning and designing will lead to excellent results. If the panel design is according to the standard and specifications, there will be less chance for hidden risks.

Part of design stage will be optimizing and simplifying the wiring design inside the panel, ensuring the location of cable trays and cables management within the panel, ensuring cable type, length, label texts and conditions, avoiding dust sources and ensuring panel light place and source.

3.2 Commissioning Stage

The second important stage where the effect of hidden risks can be minimized is during equipment commissioning. Checking hidden risks shall be part of commissioning check list, so panel lights, labels, cable organizations and avoiding different source of dusts shall be confirmed. Also, it is important to prepare risk assessment for any communication panel prior to commissioning activities.

Ignoring any factor during this stage will raise the effect of these hidden risks on equipment operations.

3.3 Maintenance Stage

Properly maintained equipment and processes are necessary to keep the facility functioning at its optimum capability [5]. Maintenance of the equipment will last for many years, so it is important to ensure that the hidden risks are not effecting the equipment. This will be achieved by checking these risks in maintenance check list and take immediate actions in case of any observations. Performing excellent maintenance will ensure the minimum effect of any hidden risks.

4 Supporting Factors for Mitigating Risks

Although the main mitigation factors for hidden risks will be done in the main three stages (Design, Commissioning and Maintenance), there are supporting factors that will help in mitigating them and these factors can be applied in any stage if not all. These factors are:

4.1 Knowledge Sharing

When risks are identified and mitigated, it is essential to share these risks with all concerned engineers. This knowledge sharing will ensure that all engineers are aware and familiar with these hidden risks so that they can deal effectively with them.

Knowledge sharing can be through short awareness, meetings, emails or during site visits. It is important to ensure the effectiveness of the knowledge shared where all hidden risks experienced and faced are understood by all.

4.2 Effective Communication

During any stage, it is important to communicate all challenges and issued faced including the hidden risks to all concern engineers. This will ensure that these risks are known by all. In addition, many risks can be avoided in future sites.

If an engineer found during maintenance stage that a series of panel lights for example are having an issue, it is important to communicate these issues to design engineers in order to take care about it. In addition, the same should be communicated to commissioning engineers in order to ensure the issue is tackled during commissioning process.

Effective communication among all engineers will ensure that risks are mitigated properly. It is also necessary to establish interface documents or Service Level Agreements (SLAs) among all concerns in order to have effective communication in terms of the duration and method.

4.3 Documentation and Records

Beside knowledge sharing and effective communication, there should be an effective process for documentation and records. This can be done through systems, emails or stored in organization's shared point that is accessed only by authorized engineers.

Documenting and recording all risks faced will be a reference for all engineers in different department. In addition, it will be useful for new engineers to learn about these risks and know the way of mitigating them in case faced at sites.

4.4 Involvement of Original Equipment Manufacturer (OEM)

Some risks might be faced for the first time or repeated many time, so it will be useful to involve Original Equipment Manufacturer (OEM) in process of mitigating these risks. This will support effective knowledge transfer to engineers and in the same time, recommendations will be suggested to mitigate risks in existing and future sites.

4.5 Conducting Risk Assessment

Risk Assessment (RA) is defined as a systematic process of identifying hazards and evaluating any associated risks within a workplace, then implementing reasonable control measures to remove or reduce them [6]. It is important to conduct risk assessments prior to commissioning process in order to identify all possible risks especially the hidden ones and suggest ways to mitigate them.

The same can be also conducted during some maintenance activities, which will help engineers to take care about these hidden risks. These assessments will be useful for audits and safety checks.

5 Conclusion

In conclusion, the following recommendations can be listed.

1) Hidden risk shall be identified in all equipment stages (Design, Commissioning and maintenance) in order to prevent failures as earliest as possible.
2) Risk Assessments and periodic audits on telecom equipment's are important tools to identify hidden risks and minimize their effect.
3) Communication panel and equipment specifications should contain points which will minimize the effect of hidden risks.
4) Importance role of risk management program owners to increase the awareness level of all risks types, plans and tools to mitigate it.
5) Supporting factors that will mitigate hidden risks to be considered in all stages.

References

1. Market Business News. https://marketbusinessnews.com/financial-glossary/risk-definition-meaning/. Accessed Feb 2021
2. Investopedia. https://www.investopedia.com/terms/u/utilities_sector.asp. Accessed Feb 2021
3. NFON. https://www.nfon.com/en/service/knowledge-base/knowledge-base-detail/tele-communication-networks. Accessed Sept 2020
4. Bakken, D.: Smart Grids, 1st ed, ch. 11, p. 254. Taylor & Francis Group, New York (2014)
5. Capehart, B.L., Turner, W.C., Kennedy, W.J.: Guide to Energy Management, 7th ed, ch. 10, p. 363. The Fairmont Press, Inc. (2011)
6. British Safety Council. https://www.britsafe.org/training-and-learning/find-the-right-course-for-you/informational-resources/risk-assessment/. Accessed Feb 2021

JomSnapBuy: Search and Buy Product with a Snap

H. K. Kee and P. S. JosephNg(✉)

Institute of Computer Science and Digital Innovation, UCSI University, UCSI Heights, 56000 Cheras, Kuala Lumpur, Malaysia
josephng@ucsiuniversity.edu.my

Abstract. With the growth of technologies, more and more products have been introduced daily to the world rapidly. There are quite several products that can bring convenience to our life but the people do not know the name, or where to buy it. Sometimes, people have to waste a lot of time finding it, especially for the younger generation as they are more tend to spend on the new product. This is the same as finding a song with its melody, neither lyrics nor song name is tough. So, this study is inspired by this thought and trying to come out with an application that can recognize the product in an image and find the store page online to buy it. Besides, the user can share the interesting product they found with others through this application. In this paper, we will discuss the image recognition technique and the state some survey & interview results collected by the Google Form.

Keywords: Image recognition · Object identifying · Machine vision · Deep learning · Machine learning · SnapBuy

1 Introduction

In machine vision, image recognition or image classification is the process of a machine to identify an object contains in a video or image [1]. This technique also frequently comes in augmented reality (AR) and allows users to access information like a 3D diagram, company logo, or text [2]. The introduction of the Convolutional Neural Network (CNN) has made the study field focus on image recognition [3]. This is a challenge for the computer to classify the image accurately as a human when provided with very limited details, in contrast, a human can quickly learn a new class of image from a few examples [4].

With the existence of deep learning, which tries to mimic human thinking, the accuracy of image recognition is increasing with the time flow [5, 6]. The image recognition that enhanced with adversarial examples training could have higher accuracy [7]. Now, it is frequently used in different fields like security surveillance, CAPTCHA security check, and driver assistant [8].

The idea of the app to identify the product on a snap is based on the image recognition concept above. The objective of this app is to help the users to save their time searching for products with an unknown name. With the help of the app, users can identify a single

M. Ma (Ed.): ICTCE 2020, LNEE 797, pp. 153–163, 2022.
https://doi.org/10.1007/978-981-16-5692-7_17

product on a snap and reach the online store for a certain product or its closest substitute, the users no longer need to spend time on the searching product with describing words on the search engine.

Literature Review. Deep Learning (DL) is a subfield to machine learning and it is introduced to break down the information with a homogenous structure like how humans will act [9, 10]. A neural network is the combination of bonded hardware or separated software that operates on a small part like a neuron in the human brain [1].

There are 2 types of neural networks: CNN-based network and ANN-based network. Artificial. Neural Network (ANN) and Convolutional Neural Network (CNN) [11, 12]. The major difference between ANN and CNN is that only the last layer in CNN is fully connected while every neuron in ANN is connected to each other [2]. CNN is specifically applied for image classification and object recognition. ANN is not suitable for image recognition because it can over-fitting easily due to the size of the image [2].

The Alpha-matting method was the simplest and most used method in image recognition [13] before the DL network structures. This method is one of the image matting, also a good method for image recognition other than the DL neural network [14]. Some models have been introduced to optimize image recognition's performance like instances-based optimization and shape-instability [15, 16]. Image completion is a method for computer replacing or filling the missing part which unable to identify [17–22].

On the other hand, unsupervised feature learning enables the machine to learn from unlabeled datasets and find the patterns that hidden in the data sets and improve the image recognition process, this learning method is now an important research topic in the AI learning field [23–25].

Few situations could affect the image recognition process like the noise and fluctuation in the brightness due to the environment change [19]. This is similar to the adversarial example in [6] and it can lead CNN to make wrong predictions.

In Koehler [20], the Automated Image Tagging and Reverse Image Search have been mentioned. Automated Image Tagging is the technique that will tag the image automatically by cloud service once the image is uploaded [26–28]. It helps to improve the efficiency of finding a certain image online with the tags. It will filter out all images that match the tag and show the result to the users. Few approaches have been proposed to support the Automated Image Tagging features to increase its efficiency and accuracy on tagging [28–30].

Next, the Reverse Image Search technology makes the possible on searching the original image using a slightly modified version of it [31]. As an example, Google Image, when the user uploads an image, it will retrieve the image information and showing the image with match result from its database. Image analysis can help to extract useful information and compute the big data [32].

There are two similar applications, Camfind and Google Lens. CamFind was developed in 2013 and the purpose is to search for an item by taking a photo of it without any typing and users can share the search result with others [33]. Google Lens was developed by Google's machine learning and introduced in 2017 by Google to help in recognizing plants, animals, numbers, and others [34–37].

Based on Table 1 we can compare the difference between these applications. Each application has a very good performance in finding the product online and list out the

Table 1. Current recognition apps features compare with SnapBuy

Features	SnapBuy	CamFind	Google lens
Object to recognize	Everything	Cannot recognize plants, animals, and faces	Everything
Web result	/	/	/
Related image	/	/	/
Shop result	/	/	/
Price comparison	/	/	/
Social profile and sharing	/	/	X

related information but there is a difference between the object to recognize. First, CamFind is more focused on socializing and cannot recognize the human, plants, or animals. It also supports users to share the interesting product they found by using CamFind on their social media via this application. Google Lens is more focused on object identifying, it can recognize not only things but also included humans, plants, and animals. Lastly, SnapBuy has both features mentioned above which are recognizing not only the things and have a social platform for users.

Problem Statement, Question and Objective. With all features provided in the app, the user could identify a single product quickly on a snap and reach the store page of a certain product or its closest substitute without any manually searching. The users no longer needed to spend time searching with describing words of the product on the internet or asking people.

This paper aims to answer the research objective via the research question in Table 2.

Table 2. Research questions

RQ1: Is this application help to save users' time on searching for a product with an unknown name?
RQ2: How to improve the trust of reliability in using this application?
RQ: How does this application carry out the process efficiently?

There are 3 research questions determined to examine the value creations.

RQ1: To find out if this application's features help users to save time on finding a product or not. The users could have more free time to do their desire stuff if this application helps them to save their time.

RQ2: To determine how to attract the users to use this application so users could feel comfortable while using this application and more tend to use this application in the future to search the product.
RQ3: To determine a way to improve its performance so it can work better for the users. If the users have a great experience with it then they could recommend this application to their friends or family.

Table 3. Research objective

RO1: To compare the time required by using this application and finding the product manually.
RO2: To improve the accuracy and time needed for identifying the product from an image.
RO3: To implement an AI to the system which helps in identifying the image with high accuracy.

Based on Table 3, this study is aimed to develop a mobile application that could identify a single product from an image with high accuracy within a short time and reach the product's or its closest substitute's store page.

RO1: To compare the time required by using this application and finding the product manually. This comparison could prove that it is more efficient in searching product with the application compare to searching it manually.
RO2: To improve the accuracy and time needed for identifying the product from an image. With high accuracy and fast process time, users could search for a product accurately within a short time. They could have a good experience with this application and recommend it to others.
RO3: To implement an AI to the system which helps in identifying the image with high accuracy. A good AI with deep-learning could enhance the image recognition feature in the application. It could learn itself from the mistake and improve itself next (Table 4).

- **H1:** This application can save time on finding a certain product with an unknown name.
  Finding a product with an unknown name needs a lot of time describing it to people or search engines, and sometimes they might not be able to understand the description and have to work more on it. With the help of an image recognition system, this application could identify the product from a snap and help users to save time by asking people and searching on search engines.
- **H2:** More users will choose this application to search for a product as it has high accuracy in identifying the product.

Table 4. Research hypothesis

H1: This application can save time on finding a certain product with an unknown name.
H2: More users will choose this application to search for a product as it has high accuracy in identifying the product.
H3: This application can identify a certain product accurately and redirect to the shop page of the product or closest substitute within a short time.

The users could have a good experience with this application as it has high accuracy and speed in identifying an unknown product. They might recommend this application to others and more users will choose to use this application.

- **H3:** This application can identify a certain product and redirect it to the shop page of the product or closest substitute within a short time.
  The application will identify the product in the image after the users take a picture of the product. When the product is identified, the application will search for the store page of the product or its closest substitute and redirect it to the page.

Value Creations. With all features provided in the app, the user could identify a single product quickly on a snap and reach the store page of a certain product or its closest substitute without any manually searching. The users no longer needed to spend time searching with describing words of the product on the internet or asking people. On the other hand, users can share the product that they feel interesting to their friend using the social profile on the app. The image recognition on this app can recognize most of the things from the image including the human, animal, plants, or things.

2 Methodology

This study will be applying a mixed research method that consists of both quantitative and qualitative research methods to obtain data more efficiently and effectively. To do so, surveys and interviews will be used during the data collection from the target audience. The data is to be collected using the following methodology which is summarized in the table below.

Based on Table 5, the research dimension will be a sequential design. The research methodology, mixed-mode is carried out by doing a random survey and interview as its primary data collection. The sequential design will illustrate as shown in Fig. 1 for the elaborations of the steps of data collection.

In quantitative data collection, the survey will be spread to popular social media to collect information and generalized the information collected. It will be followed by qualitative data collection. The qualitative data will be collected through a short

Table 5. Research methodology [38–53]

Research Dimension	Explanatory Sequential Design
Research Methodology	Mixed Mode
Research Methods	Comparative Analysis and Survey

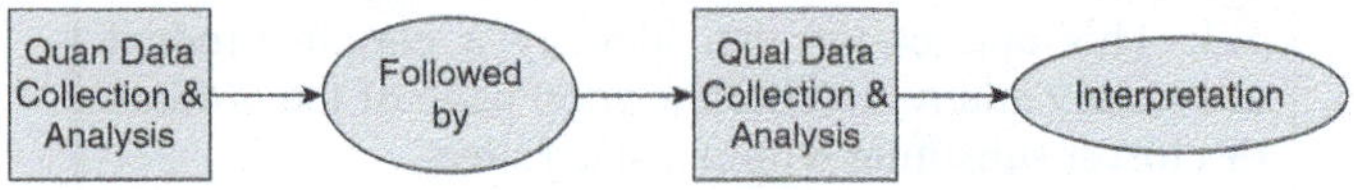

Fig. 1. Sequential design

interview with some related questions. All the information gained from the interviews will be gathered for interpretation and finally, all the information will be concluded to meet one conclusion as shown in Table 2.

3 Results and Findings

According to the data that has been found, we can see that this application is accepted by the majority of people, the result is collected by using both survey and interview questions through Google Form as data collection.

Fig. 2. Time needed to find a product manually

Figure 2 shows the result of the time needed by respondents to search for an unknown product manually in their daily life. From this result, we can see that most of the respondents need more than half an hour to search for an unknown product manually which is 43% of the total respondents are spending more than 30 min on searching. There is 1% of the respondent has mentioned that they could use several days to search for it. This result shows that most of the respondents are spending a lot of time searching when they want to buy a product but without the name and devices.

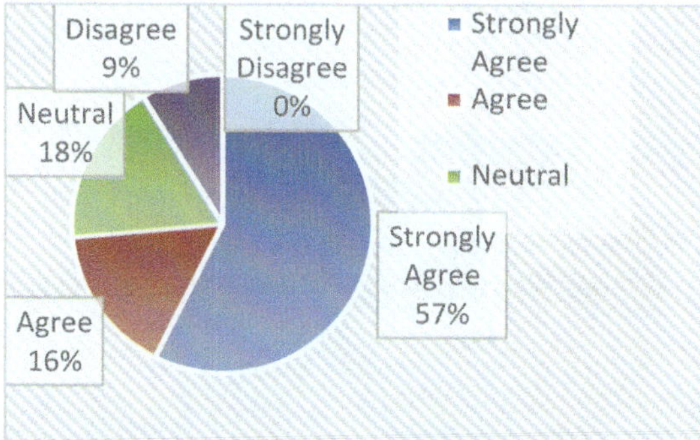

Fig. 3. Result of the thought human needs a lot of time to search for an unknown product name without the help of computers.

From the result, as shown in Fig. 3, we can see that the majority of the respondent agrees with the statement that human needs more time to search for an unknown product without the help of any devices, while there are only 9% out of the total respondents disagree about it. Therefore, we can say that a device with this app is needed for them and can make their searching process faster.

Figure 4 below shows the result of the interview question to support Fig. 3.

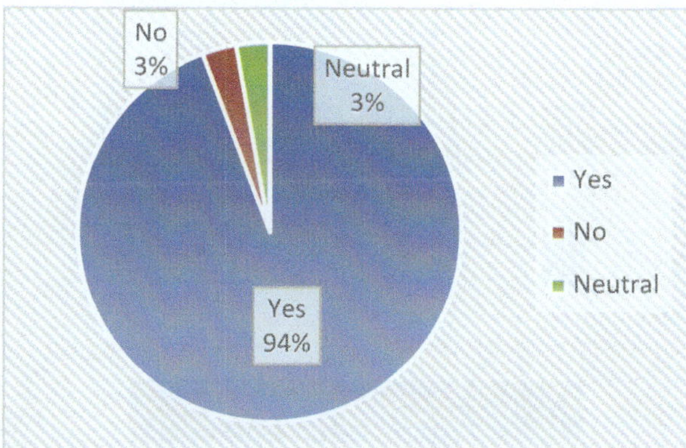

Fig. 4. Result on the users can search faster if using this app.

The chart in Fig. 4 represents the result of the interview question on the users could search faster for the unknown product faster by using this app. We can see that 94% out of the total respondents think that using this app can help to save time on searching for an unknown product while only 3% of respondents disagree with it.

From the results in Fig. 3 and Fig. 4, we can say that majority from the respondent is willing to have an app that can help them on searching the unknown product as it can help them save their times to do other works.

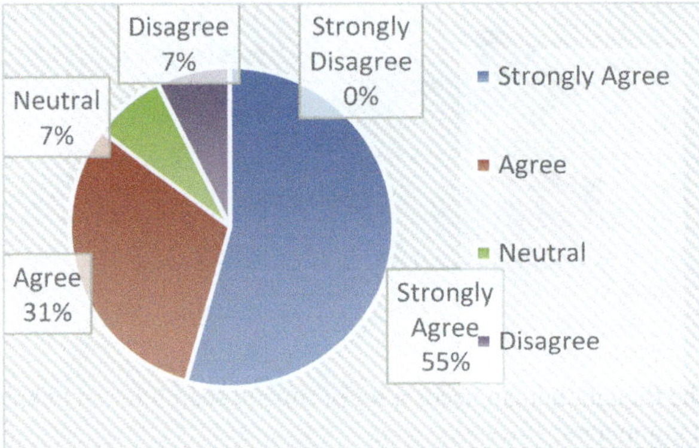

Fig. 5. Result on if the user can have a better experience with AI implemented.

Figure 5 shows the result for the question if the user can have a better experience with AI implemented in the app. 55% of respondents strongly agree that the user could have a great experience if the AI exists in the app. 31% of respondents agree with it while there are only 7% for each neutral and disagree. This result shows the majority of them would like to have the implementation of the AI in the app as the assistance and it will enhance the users' experience.

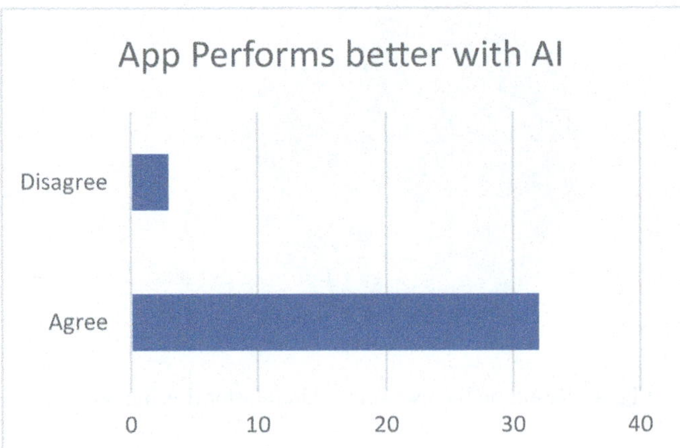

Fig. 6. Interview results on if the application works with AI can perform better.

The interview responses are group into 2 categories which are Agree and Disagree. Most of the responses are agree to the statement that application with AI performs better.

There are 91% out of the total respondents are agree with it while only 9% of respondents do not agree with it.

Based on Figs. 5 and 6, we can conclude that the appliance of AI in the app is acceptable to them. They more tend to use the app if there is an AI implementation to assist and enhance their experience while using it. Therefore, SnapBuy could have more users as there is the implementation of AI assistance on it.

4 Conclusion, Limitation, and Future Work

In conclusion, most people are spending a lot of time finding a product with an unknown name. This caused the people who do not have enough time to do their desire stuff after they have done for searching the product. On the other hand, the finding result has shown the willingness of people on the devices to help for searching the product and desire on the AI implementation in the app, therefore, we can say that SnapBuy with implemented AI assistance could be a good choice for them.

Although this application has AI implementation, and enhancement on image recognition, it might not be able to recognize some newly introduced product as the market now is introducing more and more fresh new product every day. So, this application needs a real-time update on its database in the future and the database is needed to have enough capacity to ensure that it has the data of every newly introduced product in the future.

References

1. Krishna, M., Neelima, M., Mane, H., Matcha, V.: Image classification using Deep learning. Int. J. Eng. Technol. **7**(2.7), 614–617 (2018)
2. Gogul, I., Kumar, S.: Flower species recognition system using convolution neural networks and transfer learning. In: 2017 4th International Conference on Signal Processing, Communications, and Networking (ICSCN 2017), pp. 1–6 (2017)
3. Fang, W., Zhang, F., Sheng, V.S., Ding, Y.: A method for improving CNN-based image recognition using DCGAN. Comput. Mater. Continua **57**(1), 167–178 (2018)
4. Henaff, O.J., et al.: Data-efficient image recognition with contrastive predictive coding, pp. 1–15 (2019)
5. Murase, H.: Image recognition for driver assistance in intelligent vehicles. In: 15th IAPR International Conference on Machine Vision Applications (MVA), pp. 376–381 (2017)
6. Xie, C., Tan, M., Gong, B., Wang, J., Yuille, A.L., Le, Q.V.: Adversarial examples improve image recognition, pp. 1–10 (2020)
7. Mueller, S.: Cognitive anthropomorphism of AI: how humans and computers classify images. Ergon. Des. Q. Hum. Factors Appl. 1–8 (2020)
8. Wang, J., Yu, X., Liu, Q., Yang, Z.: Research on key technologies of intelligent transportation based on image recognition and anti-fatigue driving. EURASIP J. Image Video Process. **33**, 1–13 (2019). https://doi.org/10.1186/s13640-018-0403-6
9. Cheng, F., Zhang, H., Fan, W., Harris, B.: Image recognition based on deep learning. Wirel. Peer Commun. CrossMark. **102**, 1917–1933 (2018)
10. Guo, Y., Liu, Y., Oerlemans, A., et al.: Deep learning for visual understanding: a review. Neurocomputing **187**, 27–48 (2016)

11. Wu, H., Li, Y., Miao, Z., et al.: Optimized CNN based image recognition through target region selection. Optik **156**, 772–777 (2018)
12. He, K., Zhang, X., Ren, S., et al.: Spatial pyramid pooling in deep convolutional networks for visual recognition. IEEE Trans. Pattern Anal. Mach. Intell. **37**(9), 1904–1916 (2015)
13. Wu, H., Li, Y., Miao, Z., et al.: Creative and high-quality image composition based on a new criterion. J. Vis. Commun. Image Represent. **38**, 100–114 (2016)
14. Wu, H., Li, Y., Miao, Z., et al.: A new sampling algorithm for high-quality image matting. J. Vis. Commun. Image Represent. **38**, 573–581 (2016)
15. Wu, H., Miao, Z., Wang, Y., Lin, M.: Optimized recognition with few instances based on semantic distance. Vis. Comput. **31**(4), 367–375 (2014). https://doi.org/10.1007/s00371-014-0931-8
16. Li, Y., et al.: Optimized learning instance-based image retrieval. Multimed. Tools Appl. **76**(15), 16749–16766 (2016). https://doi.org/10.1007/s11042-016-3950-9
17. Vica, V., et al.: JomImageProcessing: machine learning weight control with SnapFudo. J. Telecommun. Electron. Comput. Eng. **13**(3), 35–43 (2021)
18. Vivilyana, V., JosephNg, P.S., Shibghatullah, A.S., Eaw, H.C.: JomImage: weight control with mobile SnapFudo. In: Arai, K., Kapoor, S., Bhatia, R. (eds.) IntelliSys 2020. AISC, vol. 1252, pp. 168–180. Springer, Cham (2021). https://doi.org/10.1007/978-3-030-55190-2_13
19. Tee, F.K., et al.: JomFacial Recognition Attendance Systems. Lecture Notes in Networks and Systems (2021, forthcoming)
20. Tee, F.K., et al.: JomImage identification attendance systems. Int. J. Bus. Inf. Syst. (2021, forthcoming)
21. Vica, V., et al.: JomImage SnapFudo: control your food in a snap. In: International Conference on Engineering Technologies and Applied Sciences, Kuala Lumpur, Malaysia, pp. 1–5 (2020)
22. Liow, K.M., et al.: JomDesignLab: bringing artwork design nearer. In: IEEE International Conference on Control Systems, Computing & Engineering, Penang, Malaysia, pp. 1–6 (2020)
23. Wu, H., Miao, Z., Wang, Y., et al.: Image completion with multi-image based on entropy reduction. Neurocomputing **159**, 157–171 (2015)
24. Kang, C.M., et al.: A study on integrating penetration testing into the information security framework for Malaysian higher education institutions. In: International Symposium on Mathematical Science & Computing Research, Malaysia, pp. 156–161 (2015)
25. AdrianChin, Y.K., et al.: JomDataMining: learning behaviour affecting their academic performance, really? In: IEEE 6th International Conference on Engineering Technologies and Applied Science, Malaysia (2020)
26. AdrianChin, Y.K., et al.: JomDataMining: academic performance and learning behavior dubious relationship. Int. J. Bus. Inf. Syst. (2020, forthcoming article)
27. Wang, Y., Wang, X., Liu, W., et al.: Unsupervised local deep feature for image recognition. Inf. Sci. **351**, 67–75 (2016)
28. Nagato, T., Shibuya, H., Okamoto, H., Koezuka, T.: Machine learning technology applied to production lines: image recognition system. FUJITSU Sci. Tech. J. **53**(4), 52–58 (2017)
29. Koehler, S.: Image recognition. kitchen & bath. Business **65**(1), 34 (2018)
30. Zhu, S., Aloufi, S., El Saddik, A.: Utilizing image social clues for automated image tagging. In: 2015 IEEE International Conference on Multimedia and Expo (ICME), pp. 1–6 (2015)
31. Nwana, A., Chen, T.: Who ordered this?: exploiting implicit user tag order preferences for personalized image tagging. In: 2016 IEEE International Conference on Multimedia & Expo Workshops (ICMEW), pp. 1–6 (2016)
32. Fan, Q., Zhong, X.: A triangle voting algorithm based on double feature constraints for star sensors. Adv. Space Res. **61**(4), 1132–1142 (2018)
33. Cui, C., Shen, J., Ma, J., Lian, T.: Social tag relevance learning via ranking-oriented neighbour voting. Multimed. Tools Appl. **76**(6), 8831–8857 (2017)

34. Gao, X., Mu, T., Wang, M.: Local voting based multi-view embedding. Neurocomputing **171**, 901–909 (2016)
35. Gaillard, M., Egyed-Zsigmond, E., Granitzer, M.: CNN features for reverse image search. Document Numerique **21**(1–2), 63–90 (2018)
36. Righi, M., D'Acunto, M., Salvetti, O.: An image enhancement tool: pattern recognition image augmented resolution. Pattern Recogn. Image Anal. **26**(3), 518–523 (2016). https://doi.org/10.1134/S1054661816030160
37. Sen, Y.C., Jie, O.T.: Vision system for item recognition: case study computer peripherals. Int. J. (27), 18–21 (2019). ISSN e2600-7920
38. JosephNg, P.S., Eaw, H.C.: Making financial sense from EaaS for MSE during economic uncertainty. In: Arai, K. (ed.) FICC 2021. AISC, vol. 1363, pp. 976–989. Springer, Cham (2021). https://doi.org/10.1007/978-3-030-73100-7_67
39. JosephNg, P.S., Loh, Y.F., Eaw, H.C.: Grid computing for MSE during volatile economy. In: International Conference on Control, Automation and Systems, Busan, Korea, pp. 709–714. IEEE Explore (2020)
40. JosephNg, P.S., et al.: EaaS: available yet hidden infrastructure inside MSE. In: 5th International Conference on Network, Communication and Computing, ACM International Conference Proceeding Series, Kyoto, Japan, pp. 17–20 (2016)
41. JosephNg, P.S., Kang, C.M., Choo, P.Y., Wong, S.W., Phan, K.Y., Lim, E.H.: Beyond cloud infrastructure services in medium size manufacturing. In: International Symposium on Mathematical Sciences & Computing Research, Ipoh, Malaysia, pp. 150–155. IEEE Explore (2015)
42. Joseph, N.P.S., Choo, P.Y., Wong, S.W., Phan, K.Y., Lim, E.H.: Malaysia SME ICT during economic turbulence. In: International Conference on Information & Computer Network, Singapore, pp. 67–71 (2012)
43. Ng-Poh-Soon, J., Yin, C.P., Wan, W.S., Nazmudeen, M.S.H.: Energizing ICT infrastructure for Malaysia SME during economic turbulence. In: Student Conference on Research and Development, Cyberjaya, Malaysia, pp. 328–322. IEEE Explore (2011)
44. JosephNg, P.S.: Economic turbulence & EaaS grid computing. Int. J. Bus. Forecast. Market. Intell. (2021, forthcoming)
45. JosephNg, P.S., Eaw, H.C.: Still technology acceptance model? Reborn: exostructure as a service model. Int. J. Bus. Inf. Syst. (2021, forthcoming)
46. JosephNg, P.S.: EaaS infrastructure disruptor for MSE. Int. J. Bus. Inf. Syst. **30**(3), 373–385 (2019)
47. JosephNg, P.S.: EaaS optimization: available yet hidden information technology infrastructure inside medium size enterprises. J. Technol. Forecast. Soc. Change **132**(July), 165–173 (2018)
48. JosephNg, P.S., et al.: Exostructure services for infrastructure resources optimization. J. Telecommun. Electron. Comput. Eng. **8**(4), 65–69 (2016)
49. JosephNg, P.S., Kang, C.M.: Beyond barebone cloud infrastructure services: Stumbling competitiveness during economic turbulence. J. Sci. Technol. **24**(1), 101–121 (2016)
50. Joseph, N.P.S., Mahmood, A.K., Choo, P.Y., Wong, S.W., Phan, K.Y., Lim, E.H.: Barebone cloud IaaS: revitalization disruptive technology. Int. J. Bus. Inf. Syst. **18**(1), 107–126 (2015)
51. Joseph, N.P.S., Mahmood, A.K., Choo, P.Y., Wong, S.W., Phan, K.Y., Lim, E.H.: IaaS cloud optimization during economic turbulence for Malaysia small and medium enterprise. Int. J. Bus. Inf. Syst. **16**(2), 196–208 (2014)
52. Joseph, N.P.S., Mahmood, A.K., Choo, P.Y., Wong, S.W., Phan, K.Y., Lim, E.H.: Battles in volatile information and communication technology landscape: the Malaysia small and medium enterprise case. Int. J. Bus. Inf. Syst. **13**(2), 217–234 (2013)
53. Joseph Ng, P.S., Choo, P.Y., Wong, S.W., Phan, K.Y., Lim, E.H.: Hibernating ICT infrastructure during rainy days. J. Emerg. Trends Comput. Inf. Sci. **3**(1), 112–116 (2012)

Base Station Mobile Traffic Prediction Based on ARIMA and LSTM Model

Yige Wang(✉)

Beijing Jiaotong University, Beijing 100044, China
18711067@bjtu.edu.cn

Abstract. Two forecasting models are proposed in this study which is the Autoregressive Integrated Moving Average model (ARIMA) compared with Long Short-Term Memory (LSTM), aiming to find a model with higher accuracy of the base station mobile traffic prediction. The spatiotemporal time series is an hourly flow record of a community, unsteady and complex. In terms of ARIMA, the time series has firstly tested the stationarity by ADF method and the model parameters (p, d, q) are determined by ACF and PACF. After the model is primarily built, AIC and BIC are applied to evaluate the model. Then, LSTM models are established with three hidden layers, and the optimal model is chosen by measuring different parameter combinations. The prediction performance is evaluated by RMES, MRE, and R (correlation coefficient). With the comparison of ARIMA, the experimental results show LSTM has higher accuracy in terms of this base station traffic dataset.

Keywords: ARIMA · LSTM · Base station mobile traffic · Time series prediction

1 Introduction

With the development of mobile communication technology, 4G and 5G have brought great convenience to people and mobile traffic is growing explosively, so the traffic load of a base station is becoming increasingly significant. On the one hand, the problem of the load exceeding capacity makes the network speed pretty slow, bringing a deprived experience to users. On the other hand, due to the tide phenomenon of the base station, the number of users will dramatically decline during some periods, which is a great waste of resources and energy if the base station configuration is still operated at the carrier frequency quantity of the high-capacity period. Therefore, it is necessary to calculate the required amount of carrier frequencies according to the change of traffic and regulate it at different periods.

The parametric model is the most conventional approach used in mobile traffic prediction, including ARIMA, SARIMA, and Kalman filter [1]. Contreras [2] used ARIMA model to forecast the future electricity prices of Spain and the Californian market and maintained ARIMA has high accuracy as well as mathematical soundness. Kalpakis [3] applied the Linear Predictive Coding (LPC) cepstrum of time-series to cluster and index

M. Ma (Ed.): ICTCE 2020, LNEE 797, pp. 164–175, 2022.
https://doi.org/10.1007/978-981-16-5692-7_18

ARIMA time sequences, with desirable coefficients to demonstrate features of a dataset. Benvenuto [4] used various statistical methods to predict the trend of the prevalence of COVID-2019, and he presented the simple models like ARIMA can be effective. On the other hand, the deep learning approach drew the attention of numerous scholars and become a new trend in time series prediction. A long short-term memory network (LSTM) is increasingly more popular than deep belief networks (DBN) and recurrent neural networks (RNN) because a longer period can be calculated. H. Shao [5] took advantage of LSTM to extract features of the nonlinear traffic flow dataset. Z. Zhao [6] utilized a destination correlation matrix to show the relations of diverse links to boost the performance of LSTM. Petersen [7] combined the convolutional and LSTM method to leverage the non-static spatio-temporal correlations of the bus network. He found the complex patterns of time series cannot be captured by traditional methods.

ARIMA model is widely considered as the most efficient mathematical method in analysis and prediction due to its simpleness and interpretability. However, ARIMA does not assume the existence of any underlying model or relationships as other non-linear models, so uncertainty is essential for ARIMA model. Additionally, it is highly influenced by past values and previous error terms in prediction [4]. Although LSTM is famous for its ability to solve nonlinear and uncertain change problems, the constructure of it is overly complicated, overloading comprehension difficulty. In this study, ARIMA and LSTM models are explored and compared to clarify the superiority of each model and improve the prediction accuracy of mobile traffic flow.

2 ARIMA Methodology

2.1 Theory of ARIMA Model

ARIMA model is one of the common random time series analysis models, which is created by Box and Jenkins and is also called B-J method. The basic thought of ARIMA is to approximately recognize the change rule of the whole time series with a mathematical model, by which can more essentially understand the structure and characteristics of the time series and achieve the optimal prediction in the sense of minimum variance. (p, d, q) model (d is the order of making a stationary difference in a time series, and p and q represent the order of AR part and MA part respectively) is combined with autoregression model (AR) and moving average model (MA):

$$x_t = \varphi_0 + \sum_{i=1}^{p} \varphi_i x_{t-i} + \varepsilon_t + \sum_{j=0}^{q} \phi_j \varepsilon_{t-j} \tag{1}$$

Where φ_0 is a constant number, φ_i is the coefficient of AR part of the model, and ϕ_i is the coefficient of MA part of the model, and ε_t is the error term at time t [8];

The stability of the ARIMA model only depends on the AR part, whose homogeneous part can be transformed into the characteristic equation:

$$x^p = \alpha_1 x^{p-1} + \alpha_2 x^{p-2} + \ldots + \alpha_p \tag{2}$$

The equation is a polynomial of order p, and we can find p solutions. If the modulus lengths of these p solutions are all less than 1, then $\{x_t\}$ is stationary, or the AR model corresponding to $\{x_t\}$ is stationary. If the modulus length of k of these p solutions is equal to 1, then $\{x_t\}$ is the k order unit root process, which can be transformed into a stationary time series by k order difference. Augmented Dickey-Fuller test method is a common method to test the stationary of time series.

By calculating the autocorrelation function (ACF) and partial autocorrelation function (PACF) of the corresponding data, and comparing it with the inherent autocorrelation characteristics of the model, the suitable model parameters can be determined. The statistical related features of AR, MA, and ARIMA recognized by ACF and PACF is (Table 1):

Table 1. The statistical related features of AR, MA and ARIMA

Model	ACF	PACF
AR	Gradually decays (trailing)	Truncated after order p
MA	Truncated after order p	Gradually decays (trailing)
ARIMA	Gradually decays (trailing)	Gradually decays (trailing)

2.2 Evaluation Methods of Model Sufficiency

The evaluation methods of model performance can provide guidance to optimize the model. This study uses Akaike's Information Criterion (AIC) and Schwarz's Bayesian Information Criterion (BIC) as evaluation standards. The models with smaller AIC or BIC values are usually preferred [9].

AIC uses a function based on the likelihood to choose the best-fitted model and can deal with the overfitting situation by introducing the penalty term based on the number of free parameters. The formula of AIC is: $AIC = 2m - 2\log(L)$, where $m = p + q$ is the number of independent parameters of the model, and L represents the maximum likelihood function of the model.

BIC is closely similar to AIC. It is also based on likelihood and uses a penalty term based on the number of the estimated parameters to address the problem of over-fitting. The main difference between them is that BIC attaches more importance to the number of free parameters. The formula of BIC is: $BIC = In(m) - 2In(L)$, where $m = p + q$ is the number of independent parameters of the model, and L represents the maximum likelihood function of the model. The BIC criterion can be preferred because BIC has a larger penalty coefficient for the complexity of the model.

3 Base Station Flow Prediction Based on ARIMA

3.1 Base Station Traffic Data

Base station traffic data from the Mathorcup Big Data Challenge Competition is about changes over time of the upstream and downstream traffic in the community (each sector of the base station). The data used in this study is averaged over one hour and stored as hourly data, taken from a community from 0:00 on March 1, 2018 to 16:00 on April 18, 2018. The data has a total of 1193 observations, and then it is sorted by time order and removed for duplicate values. We use 120 samples of the leaving 1169 observations as the test set. The upstream and downstream flow of this community from March 1st to April 18th is shown in the following figure, from which we can see the trend and characteristics of the flow change (Figs. 1 and 2).

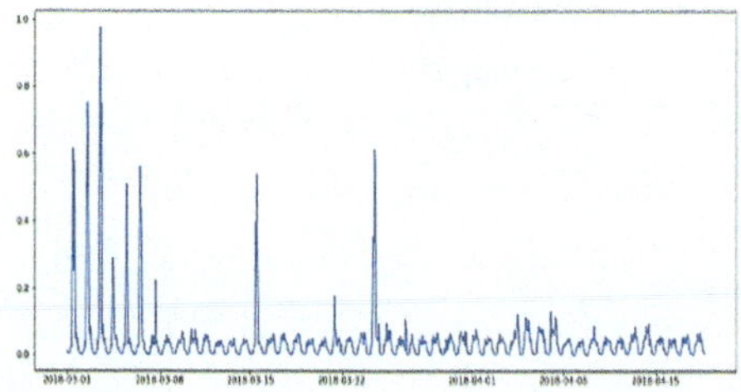

Fig. 1. The time series plot of the upstream flow

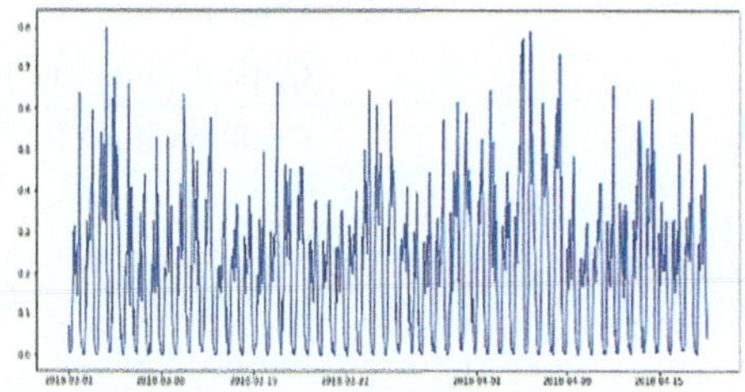

Fig. 2. The time series plot of the downstream flow

3.2 Modeling Traffic Data Using ARIMA Method

The main idea of modeling time series is: First analyze the data characteristics and use ADF method to test the stationarity of time series, which is the prerequisite for ARIMA modeling. Then, the model parameters (p, d, q) are determined by ACF and PACF. After the model is primarily built, AIC and BIC are applied to evaluate the model. Finally, optimize the model according to the parameters and make a prediction based on the existing dataset.

It can be seen from the time series plot that there are outliers taking away from the basic changing range of the upstream and downstream flow of the community so that the two-time series don't conform to stability. It is found that the time series is non-white noise series by the LBQ test, and it is statistically confirmed to be stable series by the ADF test (Table 2).

We plot the autocorrelation coefficients and the partial autocorrelation coefficients of the time series, finding the two coefficients neither gradually reduce to 0, so make a first-order difference. Test the autocorrelation coefficients and the partial autocorrelation coefficients again, and the first-order difference can be used to build ARIMA model. After tests and comparisons many times, considering the values of AIC and BIC as well as the significance of the coefficients, it is found that ARIMA (2, 1, 0) is the best choice. Therefore, ARIMA (2, 1, 0) model is built for upstream flow (Fig. 3 and Tables 3, 4).

Table 2. The test result by ADF method

	Value
Test statistic value	−4.0398
P-value	0.001216
Lags used	22
Number of observations used	1051
Critical value (1%)	−3.43659
Critical value (5%)	−2.86429
Critical value (10%)	−2.56824

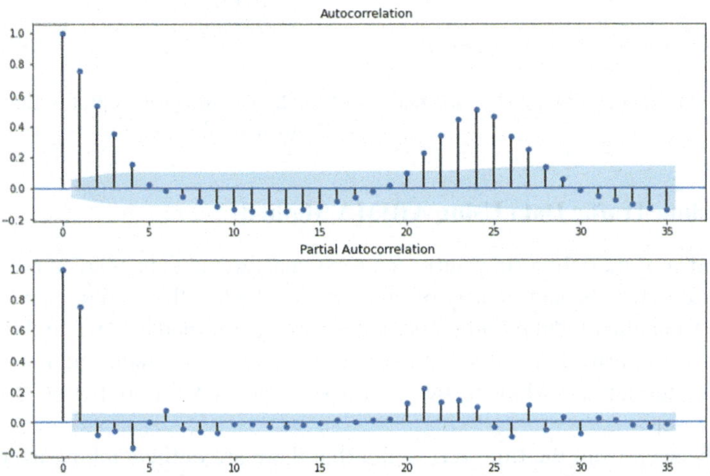

Fig. 3. The autocorrelation and partial autocorrelation of the upstream traffic

3.3 AMIRA Model Evaluation

The coefficient significance of both the upward and downward flow models is passed. Evaluated by AIC and BIC, the AIC and BIC of the upstream flow were −2835.083 and −2810.310 respectively, and the AIC and BIC of the downflow were −1273.572 and −1258.711, which are not pretty ideal in terms of the fitting effect. The figure of the comparison between predicted value and actual value also shows that there is still a gap between the fitting result and the actual result, which is due to the influence of outliers in the data. Therefore, for complex time series data, especially when the data changes irregularly and fluctuates dramatically if the outliers are not removed, the ARIMA model is powerless (Figs. 4 and 5).

Table 3. ARIMA (2, 1, 0) result of the upstream traffic

	Model	ARIMA (2, 1, 0)	Log likelihood	1422.541		
	S.D. of innovations	0.062	AIC	−2835.08		
	BIC	−2810.31	HQIC	−2825.69		
	coef	std err	z	P > |z|	[0.025	0.975]
Const	−2.99E−06	0.002	−0.002	0.999	−0.003	0.003
ar.L1.D.up quantity of flow	−0.0543	0.031	−1.762	0.078	−0.115	0.006
ar.L2.D.up quantity of flow	−0.0771	0.031	−2.506	0.012	−0.137	−0.017

Table 4. ARIMA (1, 2, 0) result of the downstream traffic

	Model	ARIMA (1, 2, 0)	Log likelihood	639.786		
	S.D. of innovations	0.131	AIC	−1273.572		
	BIC	−1258.711	HQIC	−1267.936		
	coef	Std err	z	P > |z|	[0.025	0.975]
Const	−5.15E−05	0.003	−0.02	0.984	−0.005	0.005
ar.L1.D2.down quantity of flow	−0.5737	0.025	−22.684	0	−0.623	−0.524

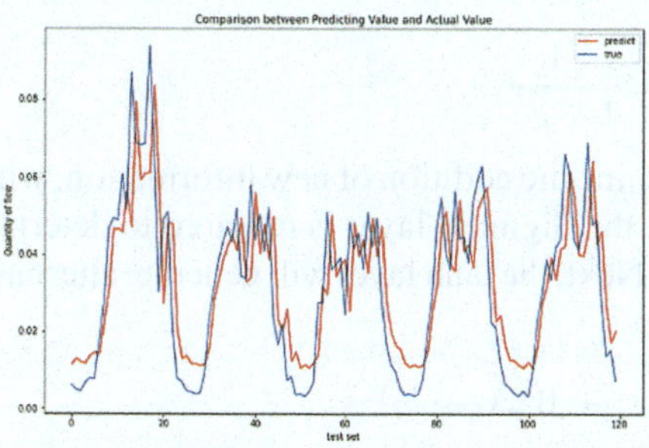

Fig. 4. Fitting performance of upstream flow compared by test set

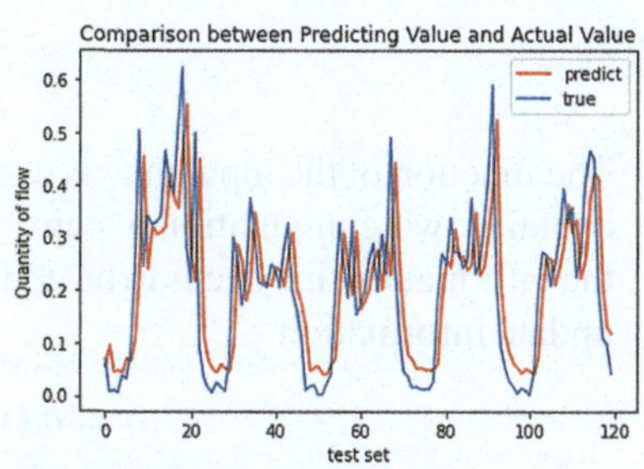

Fig. 5. Performance of downstream flow compared by test set

4 LSTM Methodology

4.1 Theory of LSTM Model

RNN is one of the neural networks that is used for training time sequences. What difference between conventional networks and RNN is that RNN allows the network internal state to memorize previous inputs, which can make an influence on the network output. Although RNN has a strong ability to model nonlinear time series, such as language, image speech recognition, and prediction, it can not process some time series with long legs. LSTM multi-layer prediction model is selected for modeling, which is derived from the recursive neural network (RNN). LSTM prediction algorithm is more suitable for complex time series, and it can solve the gradient vanishing and gradient explosion problems in the process of RNN training. There are a set of memory blocks containing one or more self-connected memory cells and gates [10] (Fig. 6).

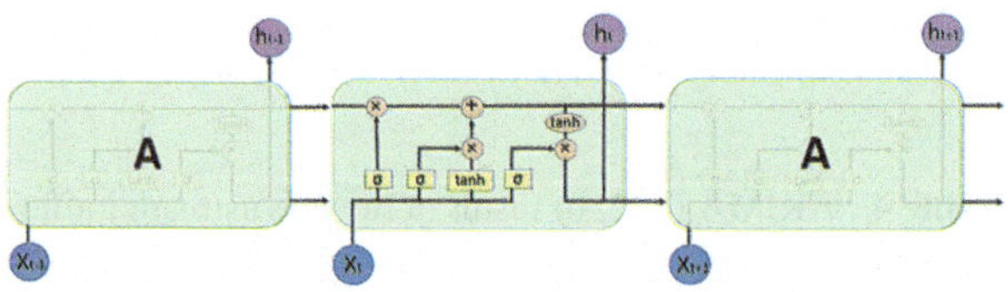

Fig. 6. The structure of LSTM memory bocks

The gate mainly realizes the control of information through the Sigmoid layer, and the output elements from Sigmoid are the real number between 0 and 1. Each LSTM memory block is composed of three gates, namely, forget gate, input gate, and output gate.

1) The function of the forget gate is to determine when the data is discarded by the cell, and then reads $\mathrm{x_t}$ and h_{t-1} as well as outputs by the Sigmoid function.

$$f_t = \sigma\left(W_{f_1} h_{t-1} + W_{f_2} x_t + b_f\right), \tag{3}$$

$$\sigma(z) = \frac{1}{1 + e^{-z}} \tag{4}$$

2) The function of the input gate is to determine the addition of new information, which contains two computational steps: First, the Sigmoid layer is required to determine the information that needs to be updated; Next, the tanh layer will generate alternative update information.

$$i_t = \sigma\left(W_{i_1} h_{t-1} + W_{i_2} x_t + b_i\right) \tag{5}$$

$$\tilde{C}_t = \tanh\left(W_{c_t} h_{t-1} + W_{c_2} x_t + b_C\right), \tag{6}$$

$$\tanh(z) = \frac{e^z - e^{-z}}{e^z + e^{-z}} \tag{7}$$

3) A new candidate value is generated by the following formula, namely, the updated cell state.

$$C_t = f_t \times C_{t-1} + i_t \times \tilde{C}_t \tag{8}$$

4) The output gate determines the output value. For the first step, the Sigmoid function is used to select which part of the cell is to be output. Then, the cell state is processed by tanh function and then multiplied by the Sigmoid output to obtain the final output value.

$$o_t = \sigma\left(W_{o_1}h_{t-1} + W_{o_2}x_t + b_o\right), \tag{9}$$

$$h_t = o_t \times \tanh(C_t) \tag{10}$$

In order to improve the accuracy of model prediction, a stack LSTM model is built by adding multiple hidden layers to the original LSTM model. The stacked LSTM model actually increases the depth of the network, and the new hidden layers recombine the learning results of the previous layers and create new dimensional representations (Fig. 7).

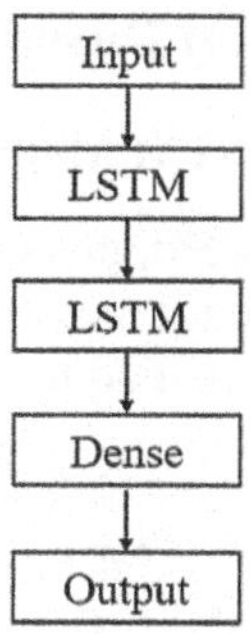

Fig. 7. The structure of stack LSTM model

4.2 Methods for Overfitting

EalyStopping and Dropout methods are used to avoid overfitting. EalyStopping refers to that in order to reduce the training model time and prevent over-fitting, stop training when the change range of loss value of training set is less than a certain value. The function of Dropout is to randomly set part of the unit to 0 at a certain probability when the model training is updated, so as to prevent the model from the dependence of on individual units and thus avoid overfitting.

Cao [11] proposed the dropout method and applied in RNN. Each time steps (inputs and outputs) at the same layers should be repeat the same dropout function, and the network units in the different layers should be applied different dropout function. The applied function of dropout in LSTM is the same as RNN. In addition, what merits attention is that the discarded units are random, whose parameters will not be updated, and all the units will work in the testing process.

4.3 Evaluation Methods of Model Sufficiency

In order to evaluate the performance of the model, Root mean square error (RMSE), which is the most frequently used measurement of prediction performance, mean relative error (MRE), which is the ratio of absolute error to actual value, and correlation coefficient (R), which is to assess the correlation between predicted values and actual values, are selected as evaluation criteria.

$$\mathrm{RMSE} = \sqrt{\frac{1}{n}\sum_{i=1}^{n}(\hat{y}_i - y_i)^2} \tag{11}$$

$$\mathrm{MRE} = \frac{1}{n}\sum_{i=1}^{n}\left|\frac{\hat{y}_i - y_i}{y_i}\right| \tag{12}$$

$$R = \frac{\mathrm{Cov}\left(\hat{Y}, Y\right)}{\sqrt{\mathrm{Var}\left(\hat{Y}\right)\mathrm{Var}(Y)}} \tag{13}$$

5 Base Station Flow Prediction Based on LSTM

5.1 Modeling Traffic Data Using LSTM Method

In order to improve the convergence speed, prevent gradient explosion and improve the calculation accuracy, the original data are normalized. This study uses the maximum and minimum normalized processing method to reduce the data to the range [0, 1].

$$X_{norm} = \frac{X - X_{min}}{X_{max} - X_{min}} \tag{14}$$

For the purpose to improve the reliability of the base station traffic prediction model, three layers of LSTM are created in this study. The first LSTM layer provides 3D output as the input to subsequent layers, and two more hidden LSTM layers are added, which can be done by setting return_sequences to True. Add Dropout after each layer and set the parameter value to 0.5. Adam was selected as the optimizer of the model with a learning rate of 0.01 and MSE as the learning objective.

In order to improve the accuracy of the model, the parameters are adjusted. We create independent variables (timesteps = 72), and at each circulation, the former 72 observations will be considered as inputs. We create dependent variables (predict_steps = 24), and at each circulation, the latter 24 observations will be considered as outputs. To extract more features, the one dimensional value of the input is converted into a vector of dim1 dimension in the first layer, we let dim1 = 128. We set dim2 = 128, the same as dim1. In the third layer, the input vector of dim2 dimensional is converted into a vector of dim3 dimensional, we let dim3 = 256. The following figure shows the change of loss MSE with the increase of epochs during the training. According to the following table, excessive value of batch_size is likely to cause local optimization but too small batch_size will dramatically extend training time, so 32 is selected as the optimal parameter (Table 5).

Table 5. Model assessment by Batch_Size

Batch_size	8	16	32	64	128
RMSE	0.106	0.103	0.097	0.097	0.097
MRE	0.424	0.282	0.259	0.330	0.343
R	0.798	0.823	0.856	0.866	0.878

5.2 LSTM Model Evaluation

The following figure is the fitting performance comparison of the predicted value and the actual value of test data set. From the figure we can judge that there is a little difference between the predicted value and the actual value, indicating that the model can ideally predict the changing trend of the future base station flow. Even in the spikes changed in the test set, (where they changed extremely), the model LSTM shows a smaller error than the ARIMA model. Therefore, although the change of base station traffic is random and irregular, the model has been able to better generalize the change regulation of traffic (Figs. 8 and 9).

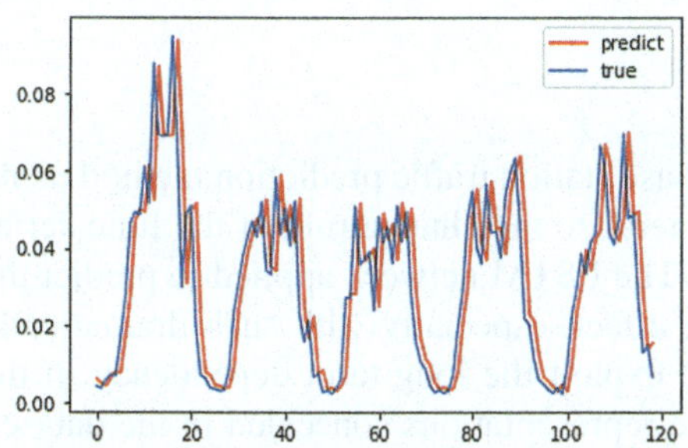

Fig. 8. Fitting performance of upstream flow compared by test set

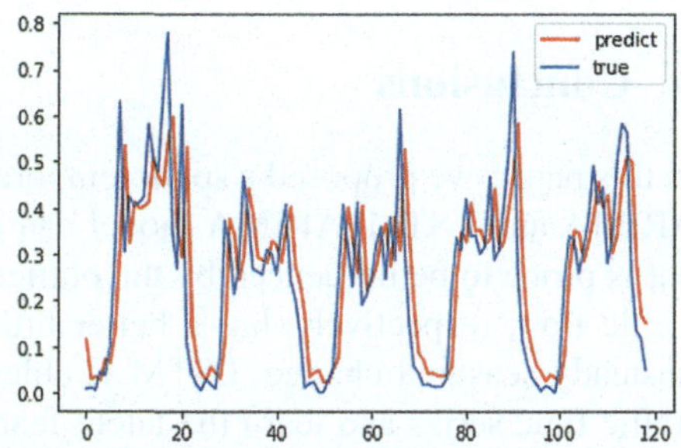

Fig. 9. Fitting performance of downstream flow compared by test set

Comparing ARIMA model with LSTM model, we find that the RMSE and MRE of LSTM model are much smaller than that of ARIMA model, indicating that LSTM has higher prediction accuracy. RMSE and MRE of the upstream and downstream flows predicted by LSTM are 0.097 and 0.259, respectively. Therefore, in the case of large-volume data, with complex time series and extreme outliers, LSTM model has a better prediction effect and can better find out change rules under the irregular dataset. The forecasting results of the flow in the next 3 days are as follows.

It can be found from the figure that the flow quantity in the next three days presents a cyclical change, which is consistent with the previous change regulation. Moreover, the change law of the predicting flow ignores the influence of the extreme values or outliers in the previous time series, so it becomes practical and stable. Besides, the overall traffic change shows an upward trend, indicating that the traffic utilization rate will increase slowly in the future, so the base station can consider increasing its traffic load (Table 6, Figs. 10 and 11).

Table 6. The comparison of ARIMA and LSTM model

	Upstream by ARIMA	Downstream by ARIMA	Upstream by LSTM	Downstream by LSTM
RMSE	0.046	0.244	0.012	0.162
MRE	3.304	8.188	0.788	1.787

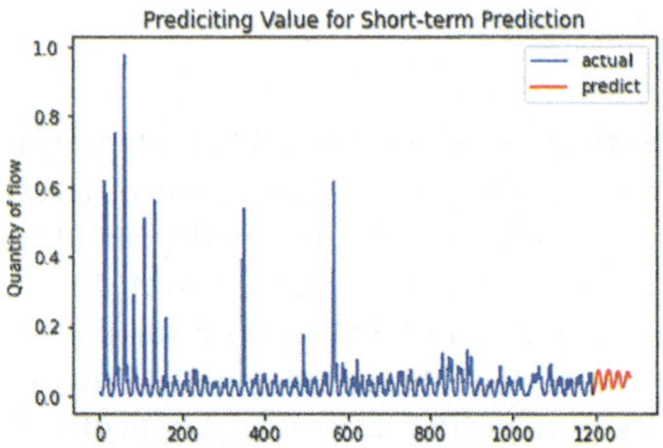

Fig. 10. Actual value and predicted value of base station upstream flow

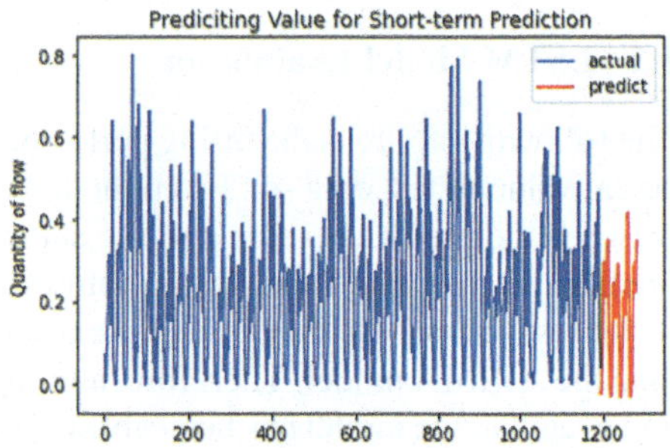

Fig. 11. Actual value and predicted value of base station downstream flow

6 Conclusions

In this paper, we proposed a spatiotemporal base station traffic prediction method using ARIMA and LSTM. ARIMA model can generalize the change rule of the time series but is prone to be influenced by the outliers. The LSTM network applied to predict the traffic flow, respectively, has a better fitting effect, especially with such dramatically unsteady seasonal change. LSTM is able to exploit the long-term dependency in the traffic time series and learn the latent feature representations concealed in the dataset, which harvests better prediction performance. We evaluate the prediction model with a test set by RMSE and MRE and compare the two models. The results present LSTM network is superior to ARIMA when the characteristics of the dataset are complex and irregular. After predicting the flow quantity in the next three days, we found the predicting series is practical and stable, verifying the success of LSTM model. And the base station can increase the traffic load due to the overall growth trend.

Although the prediction effect of LSTM neural network reaches the expected requirements, the randomness inherent in the base station traffic data still affects the accuracy of the model. In order to reduce the random fluctuation of the time series of the base station traffic, and to ensure that the traffic shows an upward trend in the long-term prediction, we will use STL combined with LSTM to decompose time series in the future.

References

1. Adebiyi, A.A., Adewumi, A.O., Ayo, C.K.: Comparison of ARIMA and artificial neural networks models for stock price prediction. J. Appl. Math. https://doi.org/10.1155/2014/614342

2. Contreras, J., Espínola, R., Nogales, F.J., Conejo, A.J.: ARIMA models to predict next-day electricity prices. IEEE Trans. Power Syst. (2003). https://doi.org/10.1109/TPWRS.2002.804943
3. Kalpakis, K., Gada, D., Puttagunta, V.: Distance measures for effective clustering of ARIMA time-series. In: Proceedings - IEEE International Conference on Data Mining, ICDM (2001). https://doi.org/10.1109/icdm.2001.989529
4. Benvenuto, D., Giovanetti, M., Vassallo, L., Angeletti, S., Ciccozzi, M.: Application of the ARIMA model on the COVID-2019 epidemic dataset. Data in Brief. (2020). https://doi.org/10.1016/j.dib.2020.105340
5. Shao, H., Soong, B.-H.: Traffic flow prediction with long short-term memory networks (LSTMs). In: Proceedings of the IEEE Region 10 Conference, Singapore, pp. 2986–2989 (2016)
6. Zhao, Z., Chen, W., Wu, X., Chen, P.C., Liu, J.: LSTM network: a deep learning approach for short-term traffic forecast. IET Intel. Transport Syst. **11**(2), 68–75 (2017)
7. Petersen, N.C., Rodrigues, F., Pereira, F.C.: Multi-output bus travel time prediction with convolutional LSTM neural network. Expert Syst. Appl. **120**, 426–435 (2019). https://doi.org/10.1016/j.eswa.2018.11.028
8. Zhang, P.G.: Time series forecasting using a hybrid ARIMA and neural network model. Neurocomputing (2003). https://doi.org/10.1016/S0925-2312(01)00702-0
9. Liu, H., Erdem, E., Shi, J.: Comprehensive evaluation of ARMA-GARCH(-M) approaches for modeling the mean and volatility of wind speed. Appl. Energy **88**(3), 724–732 (2011). https://doi.org/10.1016/j.apenergy.2010.09.028
10. Luo, X., Li, D., Yang, Y., Zhang, S.: Spatiotemporal traffic flow prediction with KNN and LSTM. J. Adv. Transp. (2019). https://doi.org/10.1155/2019/4145353
11. Cao, J., Li, Z., Li, J.: Financial time series forecasting model based on CEEMDAN and LSTM. Phys. A **519**, 127–139 (2019). https://doi.org/10.1016/j.physa.2018.11.061

Particle Swarm Optimized Optical Directional Couplers with Ultrasmall Size and Wide Bandwidth

Yuan Yang, Qichao Lu, Xin Yan, Xia Zhang(✉), and Xiaomin Ren

State Key Laboratory of Information Photonics and Optical Communications, Beijing University of Posts and Telecommunications, Beijing 100876, China
xzhang@bupt.edu.cn

Abstract. Optical directional couplers are the basic components of many optical information devices. The optical directional couplers based on silicon-on-insulator are very attractive due to the low power consumption, high transmission efficiency, and compatibility with complementary metal-oxide-semiconductor fabrication processes. However, the size of couplers designed by traditional ways could not meet the requirements of on-chip integrated optical systems. In this paper, two cross directional optical couplers on silicon-on-insulator are designed by a particle swarm optimized inverse-design method. Benefiting from the large optimization space, the footprint is as small as $2 \times 2\ \mu m^2$, which is much smaller than that of traditional devices. The simulation results show that the coupling efficiency of TE and TM directional couplers reaches 69.1% and 89.7% at 1550 nm, respectively. Both devices could operate in a broad wavelength range of 1480–1570 nm for TE and 1480–1605 nm for TM, and maintain over 60% transmission and over 20 dB crosstalk at the same time. This work may pave the way for ultra-small optical devices and on-chip photonic integrated circuits.

Keywords: Directional coupler · PSO · SOI

1 Introduction

Optical communication and signal processing have attracted much attention due to their advantages such as large communication capacity, fast processing speed and low power consumption [1–3]. Optical directional couplers are the basic components of many optical information devices such as optical filters [4], 3-dB beam splitters [5], polarization beam splitters [6, 7], integrated quantum logic gates [8], and all-optical data processing [9]. Various technologies and materials have been used to realize optical directional couplers, such as photonic wire [10], curved silicon wire [11], photonic crystal [12], topological structure [13], etc. Nowadays, photonic-integrated circuits (PICs) fabricated on silicon-on-insulator (SOI) platforms have attracted much attention due to their low power consumption, high transmission efficiency, small footprint, and compatibility with complementary metal-oxide-semiconductor (CMOS) fabrication processes. The optical

M. Ma (Ed.): ICTCE 2020, LNEE 797, pp. 176–180, 2022.
https://doi.org/10.1007/978-981-16-5692-7_19

directional couplers based on SOI are very attractive in the future optical communication and computing systems. However, the size of such devices designed by traditional ways could not meet the requirements of on-chip integrated optical systems. Here, two ultra-compact optical directional couplers with a footprint of only $2 \times 2\ \mu m^2$ on SOI are proposed. The structure of the devices is obtained by particle swarm optimization (PSO) [14] algorithm. Benefiting from the PSO algorithm's large optimization space, the feature size of our devices is much smaller than traditional devices, which makes them particularly appropriate for integration. Simulation results show that the coupling efficiency of TE/TM directional coupler reaches 69.1% and 89.7% at 1550 nm, respectively. Both devices could operate in the wavelength range 1480–1570 nm for TE and 1480–1605 nm for TM, and maintain over 60% transmission and over 20 dB crosstalk at the same time.

2 Technical Work

Here finite-difference time-domain (FDTD) [15] analysis is used for simulation, and the PSO algorithm is used to optimize the device structure.

PSO, first proposed by Dr Eberhart and Dr Kennedy, simulates the social behavior of the birds. A population of particles are initialized with random positions and velocities. Positions and velocities of each particle are updated according to Eq. (1) and (2).

$$x_{i,d} = x_{i,d} + v_{i,d} \tag{1}$$

$$v_{i,d} = \mathrm{w_n} \times \ v_{i,d} + c_1 \times rand() \times (p_{i,d} - x_i) + c_2 \times rand() \times (g_{i,d} - x_{i,d}) \tag{2}$$

where $x_{i,d}$ and $v_{i,d}$ are the i_{th} particle's positions and velocities in the d_{th} dimension of the parameter space, respectively. w_n is the inertial weight for nth iteration and determines how likely the particle stays on its old velocity. $p_{i,d}$ is individual best positions and $g_{i,d}$ is global best position. c_1 and c_2 are acceleration constants, which are equal to 2 for most applications. In our case, a large inertial weight is used to traverse most of the design space and finally a smaller inertial weight is employed for convergence. The flow chart of PSO algorithm is shown in Fig. 1.

Each device is comprised of 20 × 20 square "pixels" of size 100 nm × 100 nm, resulting in a total area of 2 μm × 2 μm [16]. Every pixel possesses two possible states: "0" or "1", indicating the pixel's material: "0" for air and "1" for silicon. The pixels are seen as particles with identical states (instead of position) and velocities. In this paper, we set fitness (objective function) by the following equation:

$$fitness = (1 - \mathrm{T_{AtoD}}) + \mathrm{T_{AtoC}}/\mathrm{T_{AtoD}} \tag{3}$$

where T_{AtoD} is the transmission from port A to D, and T_{AtoC}/T_{AtoD} is the crosstalk from C to D. Our objective is to reach the maximum directional transmission as well as the minimum crosstalk. The algorithm would stop when the fitness reaches the requirements or it meets our iteration limit. Finally, we will get the structure of the device.

The geometry of directional couplers are shown Fig. 2, (a) for TE, and (b) for TM. Each device has two input ports, A and B, and two output ports, C and D. The device

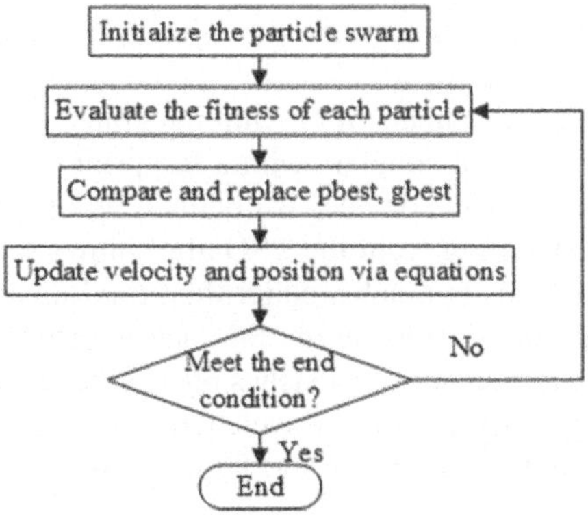

Fig.1. Flow chart of PSO algorithm.

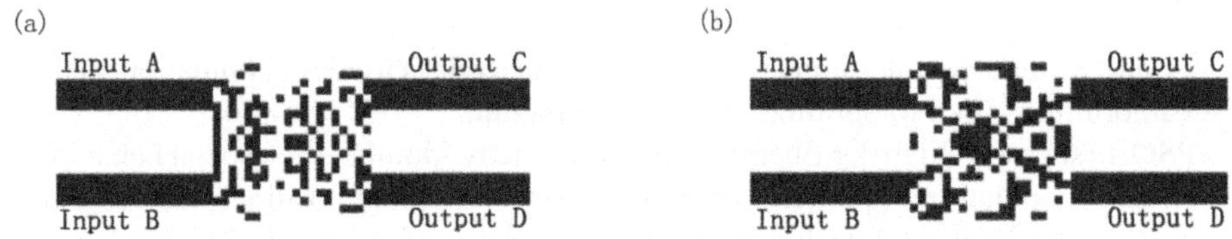

Fig. 2. Geometry of directional coupler for (a) TE, (b) TM.

could transmit the input signal from A to D as well as from B to C. The structure of the device is symmetrical up and down, so we only analyze the transmission from A to D.

To study the performance of designed directional couplers, a FDTD method is employed. Perfect match layers (PMLs) are imposed at the edges of the computational window. The silicon device layer with a permittivity of $\varepsilon_{Si} = 11.97$ is placed on top of a silica buffer layer with a permittivity of $\varepsilon_{silica} = 2.10$. Through this method, the light field at the input and output ports can be obtained, which can be used to assess the performance of the device.

As shown in Fig. 3(a) and (b), the TE/TM signal input from port A is directionally coupled to port D. Figure 3(c) and (d) show the variation curve of transmission and crosstalk with wavelength of TE and TM directional coupler, respectively. It can be seen that the transmission of the device from port A to D at 1550 nm is 69.1% for TE, and 89.7% for TM. The size of the devices is much smaller than the size of devices reported in [10–12], but the transmission is almost the same as them. This is because we set a suitable objective function, so the algorithm can find the solution in a large optimization space. Besides, the coupling efficiency exceeds 60% in the wavelength range of 1450–1570 nm for TE and 1470–1605 nm for TM. Since the algorithm's objective function not only ensures the coupled transmission, but also inhibits the crosstalk in a wide bandwidth, the crosstalk is over 20 dB in the wavelength range 1450–1650 nm for TE and 1465–1650 nm for TM.

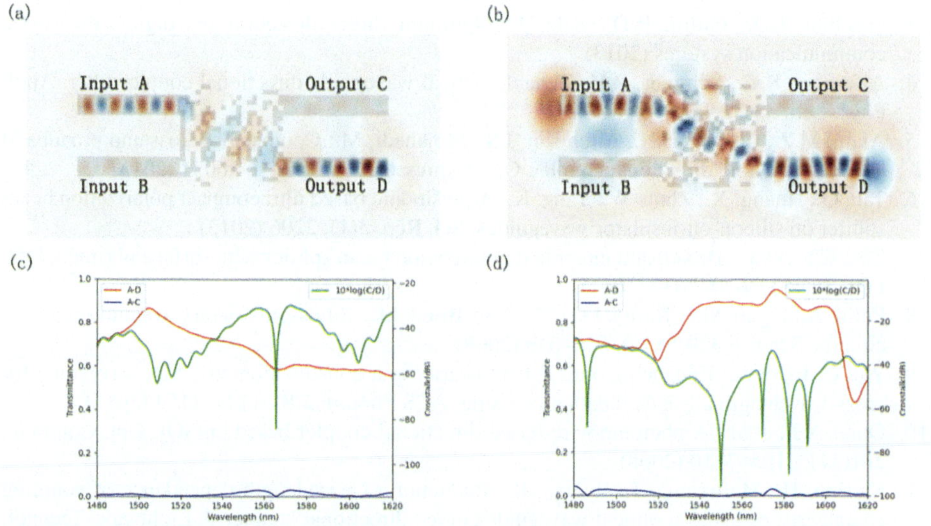

Fig. 3. (a) and (b) The FDTD simulation results of designed directional coupler for TE and TM, respectively. (c) and (d) Dependence of transmission and crosstalk on wavelength for TE and TM, respectively.

3 Conclusion

Two ultra-compact integrated optical directional couplers with a footprint of $2 \times 2\ \mu m^2$ on SOI have been proposed. The devices are designed by a nonlinear search algorithm-PSO, and exhibit extremely small size, high coupled efficiency and wide bandwidth, which may pave the way for future on-chip photonic integrated circuits. And we believe that in the future the method based on PSO can be used to design more complex directional couplers, such as increasing the number of ports, or to design other types of photonic devices for ultra-small PICs.

Acknowledgement. This work was supported by the National Natural Science Foundation of China (61935003), the National Key Research and Development Program of China (2018YFB2200104), Beijing Municipal Science and Technology Commission (Z191100004819012), the Fundamental Research Funds for the Central Universities (2018XKJC05), and the Fund of State Key Laboratory of Information Photonics and Optical Communications (Beijing University of Posts and Telecommunications), P. R. China (IPOC2019ZT07).

References

1. Soref, R.: The past, present, and future of silicon photonics. IEEE J. Sel. Top. Quant. Electron. **12**(6), 1678–1687 (2007)
2. Pospischil, A., et al.: CMOS-compatible graphene photodetector covering all optical communication bands. Nat. Photon. **7**(11), 892–896 (2013)

3. Foschini, G.J., Gitlin, R.D., Salz, J.: Optimum direct detection for digital fiber-optic communication systems (2013)
4. Alferness, R.C., Schmidt, R.V.: Tunable optical waveguide directional coupler filter. Appl. Phys. Lett. **33**(2), 161–163 (1978)
5. Alam, M.Z., Caspers, J.N., Aitchison, J.S., Mojahedi, M.: Compact low loss and broadband hybrid plasmonic directional coupler. Opt. Express **21**(13), 16029–16034 (2013)
6. Tan, Q., Huang, X., Zhou, W., Yang, K.: A plasmonic based ultracompact polarization beam splitter on silicon-on-insulator waveguides. Sci. Rep. **3**(1), 2206 (2013)
7. Zou, C.L., et al.: Broadband integrated polarization beam splitter with surface plasmon. Opt. Lett. **36**(18), 3630–3632 (2011)
8. Politi, A., Cryan, M.J., Rarity, J.G., Yu, S., O'Brien, J.L.: Silica-on-silicon waveguide quantum circuits. Science **320**(5876), 646–649 (2008)
9. Wu, C.-L., et al.: Enhancing optical nonlinearity in a nonstoichiometric SiN waveguide for cross-wavelength all-optical data processing. ACS Photon. **2**(8), 1141–1154 (2015)
10. Quan, Y.J., et al.: A photonic wire-based directional coupler based on SOI. Opt. Commun. **281**(11), 3105–3110 (2008)
11. Morino, H., Maruyama, T., Iiyama, K.: Reduction of wavelength dependence of coupling characteristics using si optical waveguide curved directional coupler. J. Lightwave Technol. **32**(12), 2188–2192 (2014)
12. Chiu, W.Y., et al.: Directional coupler formed by photonic crystal InAlGaAs nanorods. J. Lightwave Technol. **26**(5), 488–491 (2008)
13. Lu, J., Vučković, J.: Nanophotonic computational design. Opt. Express **21**(11), 13351–13367 (2013)
14. Shi, Y., Eberhart, R.: A modified particle swarm optimizer. In: The 1998 IEEE International Conference on Evolutionary Computation Proceedings, 1998. IEEE World Congress on Computational Intelligence. IEEE (1998)
15. Meep is a free and open-source, finite-difference time-domain (FDTD) software package for simulating electromagnetic systems. https://meep.readthedocs.io/en/latest/
16. Shen, B., Polson, R., Menon, R.: Integrated digital metamaterials enables ultra-compact optical diodes. Opt. Express **23**(8), 10847 (2015)

Lateral Etched Tunnel Junction Apertures for 1.3μm Vertical-Cavity Surface-Emitting Lasers

Cheng Liu[1,2,3](✉) and Huizhen Wu[3,4]

[1] School of Microelectronics, Shanghai University, Shanghai 200444, China
thomas_liucheng@shu.edu.cn
[2] Key Laboratory of Advanced Display and System Applications of Ministry of Education, Shanghai University, Shanghai 200072, China
[3] State Key Laboratory of Functional Materials for Informatics, Shanghai Institute of Microsystem and Information Technology, Chinese Academy of Sciences, Shanghai 200050, China
[4] Department of Physics, Zhejiang University, Hangzhou 310027, Zhejiang, China

Abstract. 1.3μm wafer-bonded vertical-cavity surface-emitting lasers are fabricated, which employ lateral etched p+-InAlAs/n+-InP tunnel junctions as confinement apertures. The VCSEL with a 5μm-diameter aperture exhibits CW operation with threshold current of 0.55 mA, maximum operation temperature up to 82 °C and single-transverse mode emission. The relationship between temperature-dependent characteristics and aperture sizes are discussed. The lateral etched p+-InAlAs/n+-InP tunnel junction aperture is attractive as confinement for high performance wafer-bonded long wavelength VCSELs.

Keywords: Vertical-Cavity Surface-Emitting Laser · Lateral etching · Tunnel junction

1 Introduction

In recent years, efforts have been applied to 1.3–1.55μm vertical-cavity surface-emitting lasers (VCSELs). Unlike shorter wavelength AlGaAs-based VCSELs, there are no natural oxidizable materials in an InP-based long-wavelength VCSELs to form confinement apertures. A buried tunnel junction [1–3] or lateral etched tunnel junction [4, 5] is employed to produce a low-loss thin aperture that obtains optical and electrical confinement. But up to now the lateral etched tunnel junction is only applied to the monolithic VCSELs.

In this article, lateral etched tunnel junction is used as confinement aperture in 1.3 μm wafer-bonded VCSELs.

2 Experimental

The schematic of the 1.3 μm VCSEL is illustrated in Fig. 1, a tunnel junction consisting of 20 nm n+-InP (3.0×10^{19} cm^{-3}) and 15 nm p+-InAlAs (1.0×10^{20} cm^{-3}) is located at the standing-wave node [6].

M. Ma (Ed.): ICTCE 2020, LNEE 797, pp. 181–186, 2022.
https://doi.org/10.1007/978-981-16-5692-7_20

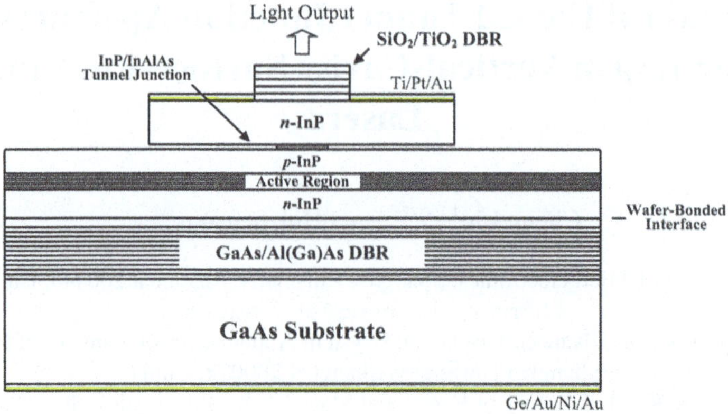

Fig.1. The schematic of the 1.3 μm VCSEL.

During fabrication, selectively lateral etching of the p-InAlAs layer is applied to obtain air-gap tunnel junction aperture. The shape of aperture is not as a circle but diamond-like square [7, 8]. The undercut etching rate of InAlAs is faster in the <001> direction than in the <011> (Fig. 2). When calculating the aperture area of the device, the diameter is defined as an average between <001> and <011> directions.

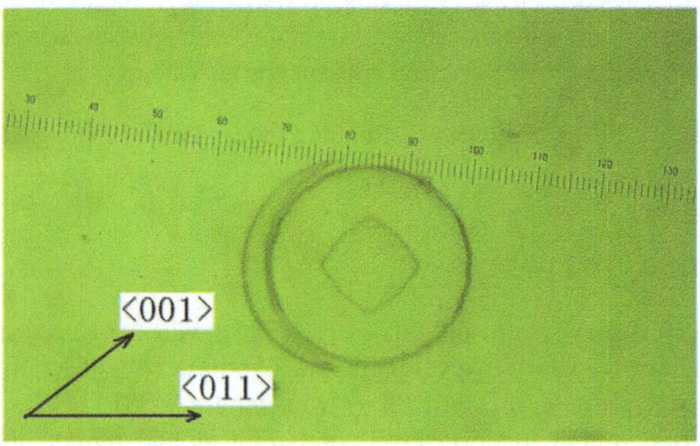

Fig. 2. Optical micrograph of lateral etched InAlAs (test structure).

Figure 3 shows a SEM picture of the InAlAs layer being etched selectively in a test ridge structure. The InP cladding layers are virtually untouched. The materials contrast between InAlAs and InP leads to a high lateral etching selectivity of about 500.

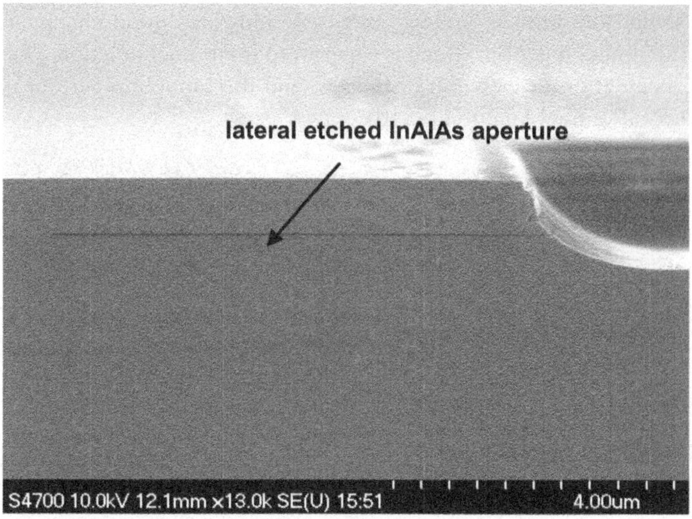

Fig. 3. Cross Section of a test structure showing the lateral etched InAlAs aperture.

3 Results and Discussion

Figure 4 shows current-voltage (I-V) characteristic at room-temperature (RT) and output optical characteristics of the 5 μm-aperture VCSELs. The threshold current (Ith) is 0.55 mA, peak optical power is 0.8 mW, and maximum operation temperature is 82 °C respectively. Single-transverse mode emission is observed with the FWHM (Full Widths at the Half Maximum) of 0.035nm at RT, which is showed in the inset of Fig. 4. The performance indicates an effective index contrast by the aperture.

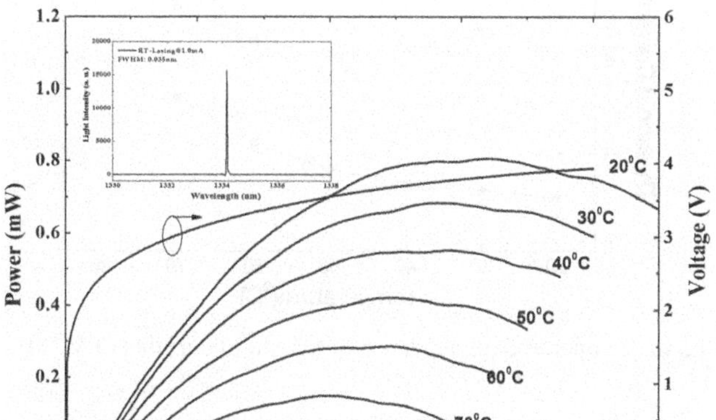

Fig. 4. L-I-V characteristics of 5μm-aperture VCSEL. **Inset:** Lasing spectrum at room temperature.

In Fig. 4, the low turn-on voltage about 2V indicates good characteristic of p+-InAlAs/n+-InP tunnel junction, which is superior to InP tunnel junction. Due to the ΔEc in Fig. 5, the depletion layer width decreases, and the tunneling current increases for p+-InAlAs/n+-InP tunnel junction [9, 10].

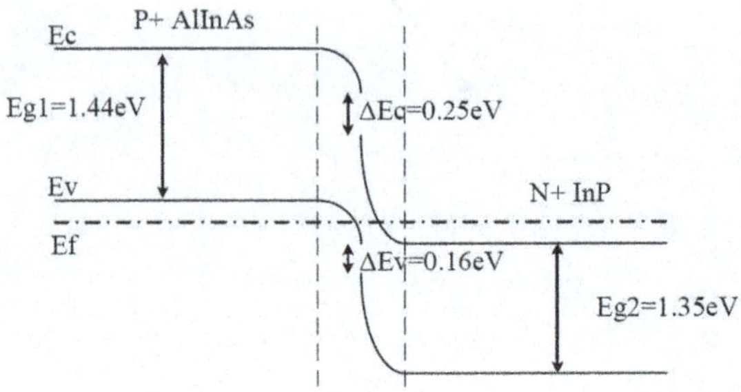

Fig. 5. Energy band diagram of p+-InAlAs/n+-InP tunnel junction.

Once the aperture is 11 μm, the VCSEL indicates almost the same turn-on voltage but multimode behaviour [6]. In order to investigate the operating mechanism of different apertures, threshold currents and external quantum efficiencies (EQEs) at a series of temperatures are showed in Fig. 6 and Fig. 7. In Fig. 6, the threshold currents continuously increase for the 5 μm-aperture, but the minimum threshold is showed at 50 °C for the 11 μm-aperture.

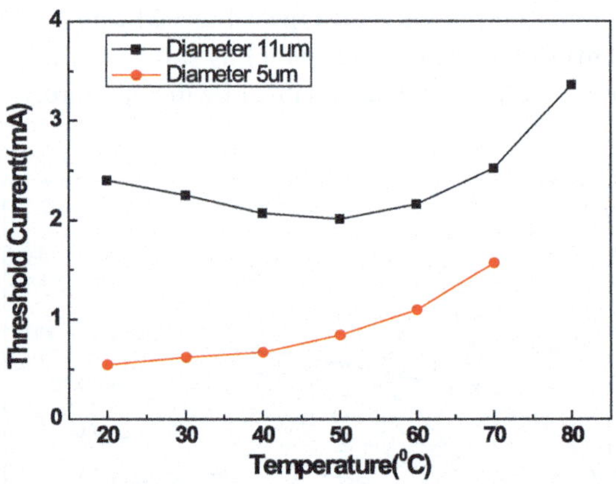

Fig. 6. Temperature dependence of threshold currents of VCSEL.

The mode-gain offsets of two devices are both −17.3 nm at room-temperature, so it will be zero at about 70 °C. But this is much different with experimental results. It comes from optical losses such as IVBA, Auger radiation, scattering and diffraction. We think that the diffraction loss is playing an important role, for the two devices have almost the same materials and processing. This point is proved from Fig. 7. The temperature-dependent EQEs have the same decreasing trends. And the 11 μm-aperture device always has larger EQE than 5 μm-aperture. So the 5 μm-aperture device has larger diffraction losses, which forms single-transverse mode emission.

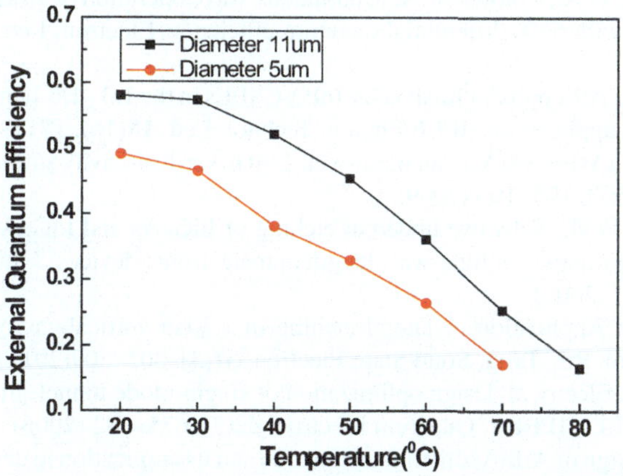

Fig. 7. Temperature dependence of external quantum efficiencies of VCSEL.

4 Conclusion

By using lateral etched p+-InAlAs/n+-InP tunnel junction apertures, CW operation of 1.3 μm VCSELs are obtained. The materials contrast leads to a high lateral etching selectivity, forming long apertures. The low turn-on voltage about 2V is obtained, due to the conduction band discontinuity of tunnel junction. Submilliamp threshold current of 0.55 mA, maximum optical power of 0.8 mW, maximum operation temperature up to 82 °C, and single-transverse mode emission have been realised. These results indicate that the lateral etched p+-InAlAs/n+-InP tunnel junction aperture is attractive as confinement for high performance wafer-bonded long wavelength VCSELs. Further improvement of device performance is needed in the future. And more characteristics such as thermal resistance of different aperture sizes, the effect of the tunnel junction position in the VCSEL structure, near field images, dynamic frequency responses are going to be investigated.

Acknowledgments. This work is supported by the State Key Development Program for Basic Research of China under grant No.2003CB314903, and the Open Fund of Key Laboratory of Advanced Display and System Applications of Ministry of Education (Shanghai University).

References

1. Muller, M., et al.: 1550-nm high-speed short-cavity VCSELs. IEEE J. Sel. Top. Quantum Electron. **17**(5), 1158–1166 (2011)

2. Spiga, S., et al.: Single-mode high-speed 1.5-μm VCSELs. J. Lightwave Technol. **35**(4), 727–733 (2017)
3. Jayaraman, V., et al.: High-power 1320-nm wafer-bonded VCSELs with tunnel junctions. IEEE Photon. Technol. Lett. **15**(11), 1495–1497 (2003)
4. Feezell, D., Buell, D.A., Coldren, L.A.: Continuous-wave operation of all-epitaxial InP-based 1.3μm VCSELs with 57% differential quantum efficiency. Electron. Lett. **41**(14), 803–804 (2005)
5. Park, M.-R., et al.: All-epitaxial InAlGaAs-InP VCSELs in the 1.3 - 1.6-μm wavelength range for CWDM band applications. IEEE Photon. Technol. Lett. **18**(16), 1717–1719 (2006)
6. Lao, Y.-F., et al.: InAsP/InGaAsP quantum-well 1.3μm vertical-cavity surface-emitting lasers. Electron. Lett. **45**(2), 105–106 (2009)
7. Pasquariello, D., et al.: Selective undercut etching of InGaAs and InGaAsP quantum wells for improved performance of long-wavelength optoelectronic devices. J. Lightwave Technol. **24**(3), 1470–1477 (2006)
8. Cheng, L., et al.: Application of lateral etching in 1.3μm vertical-cavity surface-emitting lasers (in Chinese). Res. Prog. Solid State Electron. **31**(3), 305–309(2011)
9. Mehta, M., et al.: Electrical design optimization of single-mode tunnel-junction-based long-wavelength VCSELs. IEEE J. Quantum Electron. **42**(7), 675–682 (2006)
10. Liu, C., et al.: Design of AlInAs/InP tunnel junction and its application in devices (in Chinese). Semicond. Optoelectron. **30**(5), 691–695 (2009)

Neural Network Deployment on Edge via OPC UA Protocol

Xinlei Li, Zhisheng Zhang(✉), Min Dai, and Zhangkun Shi

School of Mechanical Engineering, Southeast University, Nanjing, China
{220180326,oldbc,101010788,220190251}@seu.edu.cn

Abstract. With the rise of Industry 4.0, the transmission and processing of massive data has become a key technical problem. The OPC UA protocol is considered to be a vital solution for the unification of data transmission protocols. Edge computing services which are physically close to the site are considered to be excellent solutions to reduce transmission delay with low bandwidth occupation. However, the combination of edge computing services and OPC UA is complicated. The extra adjustment of services limits the deployment of advanced algorithm in industry field. To solve the problem, this paper proposes an OPC UA information model for neural network abstraction and instantiation. The Mobile Neural Network is introduced to the framework for rapid deployment of neural networks. Experiments proved that the online deployment, adjustment and execution of neural networks can be realized without modifying the client configuration under the OPC UA framework.

Keywords: OPC UA · Neural network · Edge computing

1 Introduction

With the increasing amount of data generated in the industrial workshop, the data transmission and processing system is becoming more and more complicated. Traditional transmission protocol cannot meet the requirement on flexible data package and analyze. The OPC Unified Architecture (OPC UA) is one of the new protocol solutions. Compared to MQTT, ROS and other protocols, OPC UA has advantages on data semantic modeling and secure transmission [1]. The wide develop language support and the tailorable features are helpful to make OPC UA be integrated into embedded devices [2]. The configurable profile and low memory usage also greatly reduces the difficulty on application deployment [3]. The release of OPC UA pub-sub model enlarges the large data capacity and greatly reduce network bandwidth overhead [4]. Because of the excellent performance, OPC UA gradually becomes a hotspot in industry data transmission field. Many manufacturers of CNC system chose to add support of OPC UA in new products. The legacy equipment also realizes OPC UA data interaction through the

---

The research work is supported by the National Natural Science Foundation of China (Grant Nos. 51775108)

M. Ma (Ed.): ICTCE 2020, LNEE 797, pp. 187–199, 2022.
https://doi.org/10.1007/978-981-16-5692-7_21

industrial Internet of things (IIOT) gateway [5]. The conversion from the original data structure to the OPC UA model can be realized easily through certain rules [6].

The transmission bandwidth and delay requirement is the second problem. It limits the application of real-time computing in industry condition [7]. Therefore, new computing structures: fog computing and edge computing are focused by researchers [8, 9]. With traditional protocol, MQTT for example, fog computing is effective in server load reduction [10]. For OPC UA, researchers concentrated on the data on fog through service aggregation. With the addition of fog computing, the redistribution of data on cloud can be canceled [11]. More importantly, by applying the edge or fog computing framework, data analysis and maintenance judgment can be realized close to the data source [12]. So the emergency message can be handled in time and the risk of data leakage is minimized. These features enable the Augmented Reality (AR) deployment in the future factory with current protocol structure [13].

In short, OPC UA and edge\fog computing are both essential for the next generation of industrial information technology. But the combination of these service is a new challenge. Currently, advanced algorithms such as neural networks are difficult to be deployed due to the existence of complicated running environment. The inconsistent model structure also adds the deployment cost [14, 15]. For computing acceleration, standalone FPGA or GPU is usually needed [16]. What is more, the transmission protocol has no native interfaces for data analyze. In order to deploy or modify the computing service, the system structure of embedded devices and remote receivers may need to be modified at the same time [17]. And the devices reboot after the program modification greatly affect the efficiency of the entire production system.

Therefore, edge computing devices in industry field need a simple neural network deployment method on common embedded chips. In order to provide a maximum device compatibility, OPC UA protocol should be added to support list. To avoid extra time cost, the deployment method should support the model update feature. To accelerate the installation progress, the dependent libraries should be divided to least and the support of popular model types should be kept.

This paper is organized as follows: Sect. 2 introduces the information model definition of neural network. Section 3 introduces the deployment method of neural network inference engine. Section 4 evaluates the performance of neural networks with OPC UA. Section 5 concludes the paper and draws some future research directions.

2 Information Model Definition

The OPC UA information model defines the organization of key parameters on the OPC UA server, including type, value, affiliation, and access authority of each data. In order to realize the deployment of the neural network, it is necessary to map all key parameters to the information model in the address space of OPC UA.

2.1 Model Definition

The model can be organized as ObjectType and VariableType. ObjectType displays the attribute characteristics of variables, including important parameters such as data type and array size. In addition, the ObjectType provides the relationship map of each variable. The client can obtain the changed information of the model in time through the event subscription function of OPC UA. VariableType can also display the properties of variables, but as a data container, it cannot show the organizational relationship between different variables. For neural networks, variables of models have different quantities and priorities. Therefore, it is more appropriate to use ObjectType.

For unified management of neural network, a template called NeuralNetworkObjectType is defined. As shown in Fig. 1, it is a subtype of BaseObjectType. Nodes are defined for data management as followed.

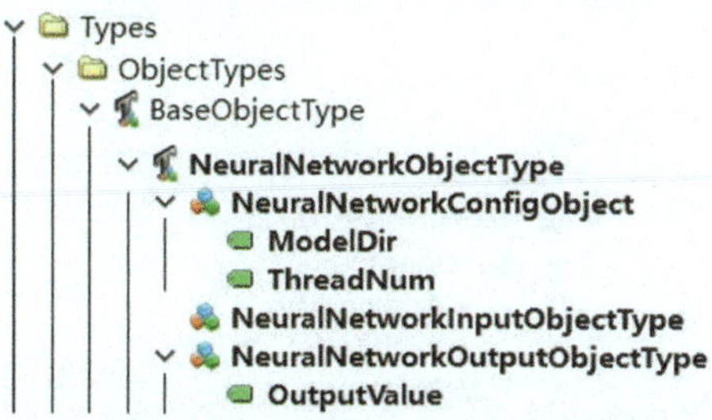

Fig. 1. Neural network object type model.

1) NeuralNetworkConfigObjectType

NeuralNetworkConfigObjectType is defined to store the operating settings of the neural network. At least, it has 2 sub-variables: ModelDir, ThreadNum.

- ModelDir represents the location of the model. For devices with a file system, this variable is the path of the file. For devices without a file system, this variable is a pointer to the program location in the storage device. The property settings of variables are readable and writable, so users own the right of remote model switching.
- ThreadNum represents the expected number of threads that the program runs with. If the inference framework supports multi-threading, it can improve the efficiency of inference and reduce processing delay. Since the embedded device may only have one computing core, only single-threaded work is allowed by default.

2) NeuralNetworkInputObjectType

NeuralNetworkInputObjectType is used to store the production data which will be imported into neural network. The variables under this node change according to the specific production process. This node is empty in the template.

3) NeuralNetworkOutputObjectType

NeuralNetworkOutputObjectType is used to store the output of the neural network. One neural network has only one output, so there is only one variable OutputValue under this node.

2.2 Model Instantiation

As shown in Fig. 2, the NeuralNetworks object is established under the Objects folder of the system. This unified entry point offers the access to manage different neural network deployments centrally.

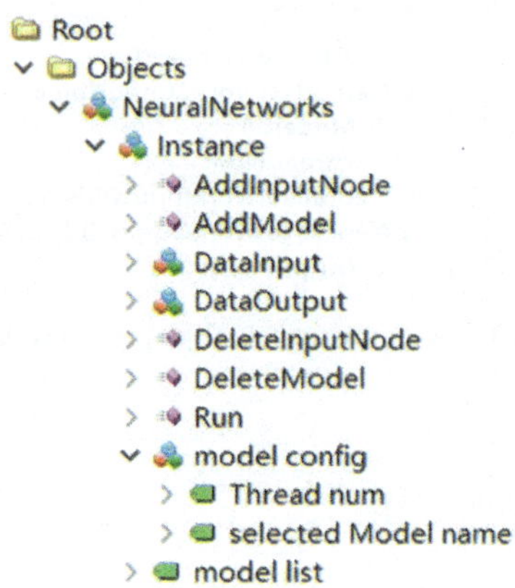

Fig. 2. Neural network instance.

1) Instance

The instance of NeuralNetworkObject under the NeuralNetworks object is created as required. Additional sub-objects can also be added for management, such as: status analysis, quality inspection, etc.

2) Model list

A ModelList variable node is created under the instance. This variable node displays the uniquely identifiers of all neural network models on device, so the administrator can decide if a update of model is required.

3) Method

For each NeuralNetworksObject instance, 5 methods are added to realize remote management of models and data.

- The AddModel() method is used to send the model to the device. By adopting the binary sending mode of ByteString, the file transmission of any format and any size within the data stream transmitting capability can be realized.
- The DeleteModel() method is used to delete the models on the device. Because of the transmission errors and the incorrect inputs, the unexpected deletion of important models may appear and lead to serious consequences. To avoid this, two input parameters about the same identifiers should be send to the server. The server can decide the next action based on the comparison result of the 2 parameters
- The AddInputNode() method is used to add the production variables into the NeuralNetworkInput object. The aggregate production variables of interest may have hundreds of nodes in the OPC UA server, so the direct copy of all data may lead to the significant increase in memory occupation. To minimize the cost of memory, the references of node are created by this method. These links clarify the affiliation of variables and no new data is created under the input object.
- The DeleteInputNode() method is used to remove the useless input variables of neural network. Since there are no copied nodes, only the references of the corresponding node need to be deleted.
- The Run() method is used to start the neural network inference. The parameters of neural network operation have been defined in the corresponding NeuralNetworksConfig, so no additional parameters are required.

2.3 Program Activation

There exists a large number of mathematical operators and parameters in neural networks. Inevitably, the execution of operations can lead to a non-negligible processor occupation and a large amount of memory consumption. For embedded systems with tight resources, neural networks are not suitable for continuous and uninterrupted operation. A suitable trigger is needed to wake up the neural network in appropriate condition.

For some time-related applications, timing sensor sampling for example, timing tasks are good triggers. By adding a timer callback function and registering the timing interval time in the OPC UA model, the running state of the neural network can be easily controlled.

For some event-related applications, such as the pulse triggered sensor sampling, event triggers are more appropriate. By adding the startup code to the callback function of the key variable, the neural network calculation can be started immediately after the value changes.

3 Inference Engine Deployment

The parameter definition, operation startup, variable input and output, model management of the neural network can be realized through the OPC UA framework. But for embedded devices, there are many dependencies required for the execution of the popular neural network models, TensorFlow, Caffe for example. To solve this problem, the open source Mobile Neural Network (MNN) is introduced into the framework.

3.1 Mobile Neural Network Instruction

MNN is an efficient and lightweight deep learning framework. With less than 1 MB of executable file size increments, inference and training of mainstream models can be realized without relying on third-party libraries [8].

MNN has a simple and easy-to-use API interface. Only four steps are needed to realize the complete operation of the neural network: create a session, input data, run the session and obtain output.

3.2 Interface Connection

MNN framework and OPC UA server use different strategies to organize data. No native interfaces exist on OPC UA server to control MNN framework. It is necessary to establish a middle layer to connect the two platforms.

1) Single variable

For a single variable, such as the number of threads in the execution of the neural network, the read value function of the OPC UA server can be directly used to obtain the corresponding UA_Variant value. As shown in Table 1, the returned UA_Variant structure contains information which can be easily transferred into the MNN style. Because of the semantic modeling of OPC UA, developers can easily understand the physical meaning of variables. One problem is that the callback function of OPC UA is static, and multiple instances may share the same callback function. The actual caller should be determined before reading the value. OPC UA offers an entry about the context parameter, and the parent node is accessible through it. Since the methods are located in the root directory of the instance when the model is defined, the parent object is the container of the desired variable and all identifiers locate there.

Table 1. Struct of UA_Variant

Attribute	Description
type	The data type description
arrayLength	The number of elements in the data array
data	Points to the scalar or array data
arrayDimensionsSize	The number of dimensions
arrayDimensions	The length of each dimension

2) Multiple variables

For multiple variables that need to be combined, such as the input variables of a neural network, data needs to be integrated and reorganized. The tensor which supports complex data structures is the common tool in data processing of neural network. But in engineering applications, most of the data is described as time sequence, which is usually no more than 2 dimensions. In order to reduce the difficulty in data organization and simplify the operation process, a 2-dimensions tensor, that is, a matrix, is used to transfer multiple variables in combination.

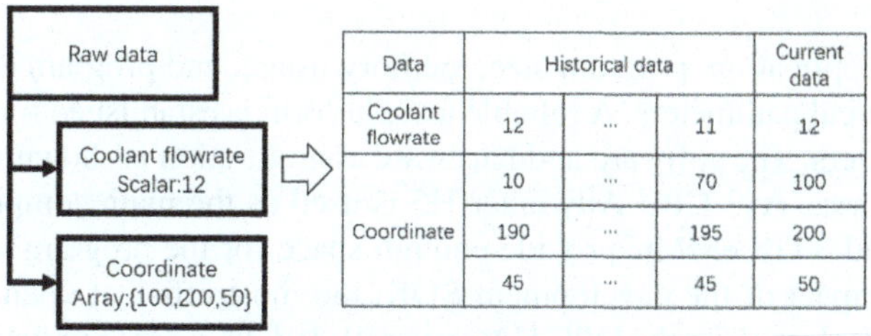

Fig. 3. Data organize

For example, as shown in Fig. 3, a milling machine has a data object in OPC UA server. User want to import the coolant flowrate and the coordinate of nozzle into neural network to monitor potential cooling system fault. For 0-dimension data, that is, the coolant flowrate, it is directly arranged in a row in time order. The length of historical data depends on the sampling window. For one-dimensional data, that is, the coordinate, a column is built in the order of the array first. Rows are filled in the same way as 0-dimensional data. After the data reconstruction, universal preprocessing can be applied to the matrix if needed. This matrix can also be directly input into characteristic recognition neural network as a "monochrome picture".

3) Function interfaces

For function interfaces, such as the establishment and execution of neural network sessions, the connection is made by adding callback functions. Due to the inherent complexity, the creation of neural networks will consume a long time, and the maintenance of sessions will consume more memory. In order to balance the cost of time and space, the neural network should be allocated according to application scenarios.

For scenarios where the time interval of neural network execution is long, the function interface of creation, input, execution, and output can be encapsulated into a callback function. When the execution conditions are met, the system directly allocates a large amount of resources for a single neural network inference. The resources are released immediately after completion. This method can minimize the memory occupation in idle state and facilitate the runtime replacement and update of the neural network. However, the allocation of large blocks of memory and the establishment of data structures will inevitably bring considerable delays.

For scenarios where the time interval of the neural network execution is short, the session creation function of the neural network can be executed simultaneously with the OPC UA initialization function. The function interface of input, execution, and output can be encapsulated in a callback function. In this way, there exists a copy of the neural network data in the memory. The network can be processed immediately with the dataflow. Obviously, a large number of neural network operators occupies a huge memory block. If the neural network update needs to be performed, a lock mechanism should also be designed to prevent the program from accessing the wrong address.

4 Evaluation

For embedded application, program size, memory usage, and program execution delay are the most critical parameters. A reliable test platform is established for the evaluation of these parameters. The software and hardware configuration is shown in Table 2. The 4-core 64-bit cortex-A53 CPU Allwinner H5 is used as the main computing unit. 512 MB memory and 8 GB eMMC provide enough space for the program to run. In order to reduce the impact of the development SDK, the most efficient open source library open62541 was selected for the OPC UA server [9, 10]. In order to reduce the switching time within the operating system, a real-time patch was additionally installed on the Linux system. All experiments use 2 threads without GPU accelerations.

Table 2. Testing setup

Name	Description
CPU	Allwinner H5@1 GHz
Memory	512 MB DDR3
Storage	8 GB eMMC
OPC UA SDK	Github Open62541 master 2020-02-29
MNN SDK	Github MNN master 2020-07-09
Linux core	5.4.26-rt17
File system	Debian 9.12
Compiler	Linaro GCC 7.4-2019.02

4.1 Program Volume

In order to evaluate the program volume increment, two different code connection methods: dynamic linking and static linking, were applied to the same program.

As shown in Fig. 4, after adding the neural network instance and MNN framework, the overall program has increased by 2.39 MB with the method of dynamic library linking. If the static linking is adopted and the MNN frame is tailored when designing the program, the program only increases by 700 KB. In contrast, under the Debian system, the dependency package of Caffe program exceeds 1 MB.

Obviously, by adopting a lightweight MNN framework and appropriate function tailoring, neural network deployment can be achieved with a small amount of storage space occupation. But for projects that use more neural networks and neural networks are designed in multiple applications, similar MNN framework code fragments will appear in multiple programs. At that time, the dynamic link mode may take up less space.

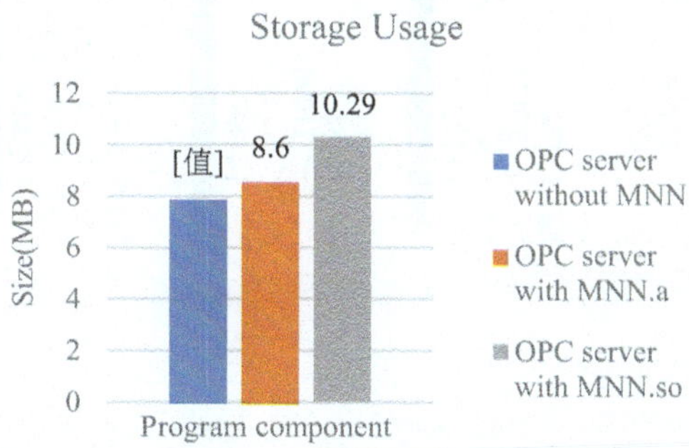

Fig. 4. Storage usage.

4.2 Memory Usage

The memory management mechanism of Linux system is complicated. Private memory and shared memory exist during program execution. Different evaluation indicators may give different results. In this experiment, the writeable/private return value of the Linux "pmap" command is used as the representative parameter.

In order to evaluate the influence of the neural network model on the overall memory usage, a common lightweight neural network was imported into the program through OPC UA. The lightweight classification/recognition network "mobilenet" was selected as the test network in this experiment. The module was send to the OPC server at runtime. To represent for the massive production data, a node of a 360*480 monochrome picture was created as the input data. It could be seen as 480 points of data source with a sampling window of 360 points. The classification results of the probability of 10 working states were exported to the neural network output node.

In order to evaluate the influence of the amount of production data on the entire project memory, additional 100/1000/10000 variables were added to the OPC UA framework. The input data of the neural network remained unchanged, so the inference action used the same size of memory.

In order to evaluate the influence of MNN function allocation strategies on memory usage, two methods: dynamic loading and static loading, were used to start MNN sessions.

The whole experiment process was repeated many times and the highest memory usage was taken as the basis for evaluation. The experimental results are shown in Fig. 5.

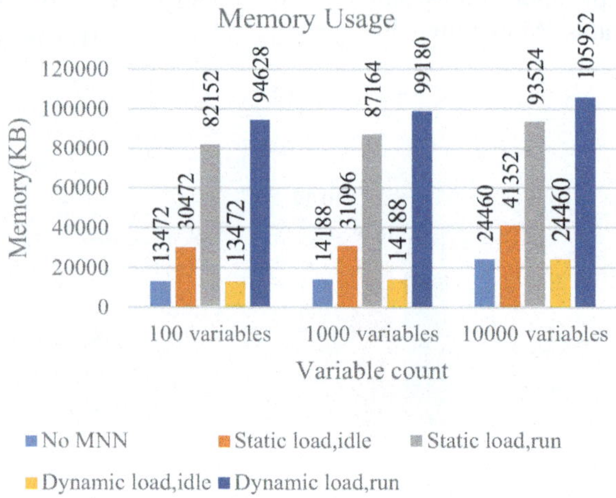

Fig. 5. Memory usage

1) Neural network model

Due to the inherent complexity of neural networks, simply loading the model will take up a lot of memory. For the "mobilenet" network, the model file volume of nearly 17 MB basically matches the memory footprint. In addition, due to the use of complex operators, a large amount of temporary space needs to be allocated to store intermediate results when performing inference. Obviously, the memory usage during inference execution will be several times than that of idle state.

2) Production data

Due to the increase in monitored variables, the memory occupation of the program will inevitably increase correspondingly. Because the namespaces under the OPC UA framework are independent of each other, the increase of variables does not affect the MNN framework. The increase in memory only related to the number of variables regardless of whether the neural network is running or not.

3) Function allocation strategies

If the neural network is statically loaded, the memory space with approximately the size of the neural network file is opened up to maintain all the necessary parameters. It leads to high memory usage throughout the entire progress. If dynamic loading is adopted, there is basically no additional load in the idle state except for the occupation of OPC UA itself. In the case of dynamic loading, the program needs more space to perform operations such as model loading and space allocation. Therefore, the maximum memory when the program is running will be significantly higher than the maximum memory during static loading.

4.3 Time Consumption

Dynamic loading and static loading not only affect the memory usage, but also affect the global time of the neural network inference. In order to test the specific differences between the two loading methods, the same "mobilenet" network was used to count the execution time of the entire neural network inference callback function. In order to evaluate the overall operating performance, the same program was also tested on a common PC.

The program run several times and the average value was taken as the basis for evaluation. The result is shown in Fig. 6.

Obviously, the use of static loading can significantly reduce the time for influence. As the steps of file loading and large memory allocation are reduced, nearly 15% performance improvement can be obtained on the embedded platform.

In addition, due to the disadvantage of computing frequency, the inference time of the embedded platform is much longer than that of the PC platform. For some scenarios with loose delay requirements, such as temperature monitoring, this inference speed can basically meet the requirements. However, for rapidly changing signals, such as motor current, GPU acceleration or the use of a better performance CPU is also required. In some case, the fog computing is a better choice.

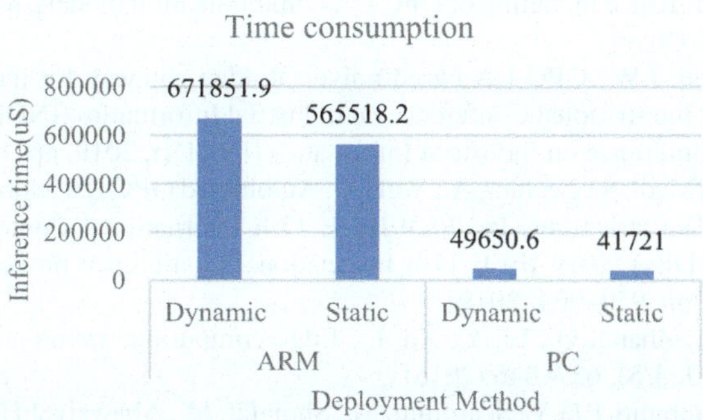

Fig. 6. Time consumption.

5 Conclusion

This paper proposes a neural network address space model based on the standard OPC UA address space, and realizes the address mapping of the important parameters of the neural network under the OPC UA framework. Under the condition of only updating the program on the OPC UA server, the neural network data view and neural network execution via the general OPC UA client are realized. Because of the existence of neural network variety and embedding complexity, this paper introduces the MNN framework as a platform to solve deployment problems. Experimental data shows that with a small amount of program storage space and a certain amount of memory, the neural network can implement flexible deployment with different focuses under the edge computing framework.

For future development, GPU acceleration is a good way to improve the inference performance. The configuration information can be added to the NeuralNetworkConfigObjectType. With the help of heterogeneous computing, the neural network can be used in more OPC UA applications.

References

1. Profanter, S., Tekat, A., Dorofeev, K., Rickert, M., Knoll, A.: OPC UA versus ROS, DDS, and MQTT: performance evaluation of industry 4.0 protocols, Melbourne, VIC, Australia, 2019, pp. 955–962 (2019)
2. Ren, H., Liu, Y., Wang, H.: Research on communication method of OPC UA client based on ARM. In: 2019 IEEE/ACIS 18th International Conference on Computer and Information Science (ICIS), 2019 IEEE/ACIS 18th International Conference on Computer and Information Science (ICIS), 2019, pp. 52–56 (2019)
3. Imtiaz, J., Jasperneite, J.: Scalability of OPC UA down to the chip level enables internet of things. In: IEEE International Conference on Industrial Informatics INDIN, 2013, pp. 500–505 (2013)
4. Burger, A., Koziolek, H., Rückert, J., Platenius-Mohr, M., Stomberg, G.: Bottleneck identification and performance modeling of OPC UA communication models, Mumbai, India, 2019, pp. 231–242 (2019)
5. Park, H.M., Jeon, J.W.: OPC UA based universal edge gateway for legacy equipment. In: 2019 IEEE 17th International Conference on Industrial Informatics (INDIN), 2019 IEEE 17th International Conference on Industrial Informatics (INDIN), 2019, pp. 1002–1007 (2019)
6. von Arnim, C., Friedl, S., Lechler, A., Verl, A.: Automated OPC UA address space generation from existing data structures. In: 2019 IEEE 17th International Conference on Industrial Informatics (INDIN), 2019 IEEE 17th International Conference on Industrial Informatics (INDIN), 2019, pp. 959–964 (2019)
7. Shi, W., Cao, J., Zhang, Q., Li, Y., Xu, L.: Edge computing: vision and challenges. IEEE Internet Things J. **3**(5), 637–646 (2016)
8. Baccarelli, E., Naranjo, P.G.V., Scarpiniti, M., Shojafar, M., Abawajy, J.H.: Fog of everything: energy-efficient networked computing architectures, research challenges, and a case study. IEEE access **5**, 9882–9910 (2017)
9. Satyanarayanan, M.: The emergence of edge computing. Computer **50**(1), 30–39 (2017)
10. Peralta, G., Iglesias-Urkia, M., Barcelo, M., Gomez, R., Moran, A., Bilbao, J.: Fog computing based efficient IoT scheme for the industry 4.0. In: Electronics, Control, Measurement, Signals and their Application to Mechatronics (ECMSM) (2017)
11. Crespi, R.S., Armentia, A., Sarachaga, I., Casquero, O., Perez, F., Marcos, M.: OPC UA Aggregation Server in the Fog, pp. 1256–1260 (2019)
12. Ashjaei, M., Bengtsson, M.: Enhancing smart maintenance management using fog computing technology. In: Industrial Engineering and Engineering Management (IEEM), IEEE International Conference on 2017, pp. 1561–1565. IEEE (2017)
13. Langfinger, M., Schneider, M., Stricker, D., Schotten, H.D.: Addressing security challenges in industrial augmented reality systems. In: Industrial Informatics (INDIN), 2017 IEEE 15th International Conference on July 2017, pp. 299–304. IEEE (July 2017)
14. Mulfari, D., Palla, A., Fanucci, L.: Embedded Systems and TensorFlow Frameworks as Assistive Technology Solutions, 2017, pp. 396–400 (2017)

15. Danopoulos, D., Kachris, C., Soudris, D.: Acceleration of image classification with Caffe framework using FPGA, Thessaloniki, Greece, 2018, pp. 1–4 (2018)
16. Lu, Y., Chen, Y., Li, T., Cai, R., Gong, X.: Convolutional neural network construction method for embedded FPGAs oriented edge computing. Jisuanji Yanjiu yu Fazhan/Computer Research and Development, 01 Jan 2018, vol. 55, pp. 551–562 (2018)
17. Kumaraguru, S., Kulvatunyou, B., Morris, K.C.: Integrating real-time analytics and continuous performance management in smart manufacturing systems. In: Grabot, B., Vallespir, B., Gomes, S., Bouras, A., Kiritsis, D. (eds.) APMS 2014. IAICT, vol. 440, pp. 175–182. Springer, Heidelberg (2014). https://doi.org/10.1007/978-3-662-44733-8_22
18. MNN repository. https://github.com/alibaba/MNN
19. Cenedese, A., Frodella, M., Tramarin, F., Vitturi, S.: Comparative assessment of different OPC UA open-source stacks for embedded systems, 2019, pp. 1127–1134 (2019)
20. "Open62541 repository". https://github.com/open62541/open62541

Author Index

M. Ma (Ed.): ICTCE 2020, LNEE 797, pp. 201–202, 2022.
https://doi.org/10.1007/978-981-16-5692-7

The manufacturer's authorised representative in the EU is Springer Nature Customer Service Centre GmbH, Europaplatz 3, 69115 Heidelberg, Germany. If you have any concerns regarding our products, please contact ProductSafety@springernature.com

Printed and bound by CPI Group (UK) Ltd, Croydon, CR0 4YY
30/09/2026
02225287-0005